Lisa LeFlore

THE WORLD OF DINOSAURS

THE WORLD OF DINOSAURS

MICHAEL TWEEDIE

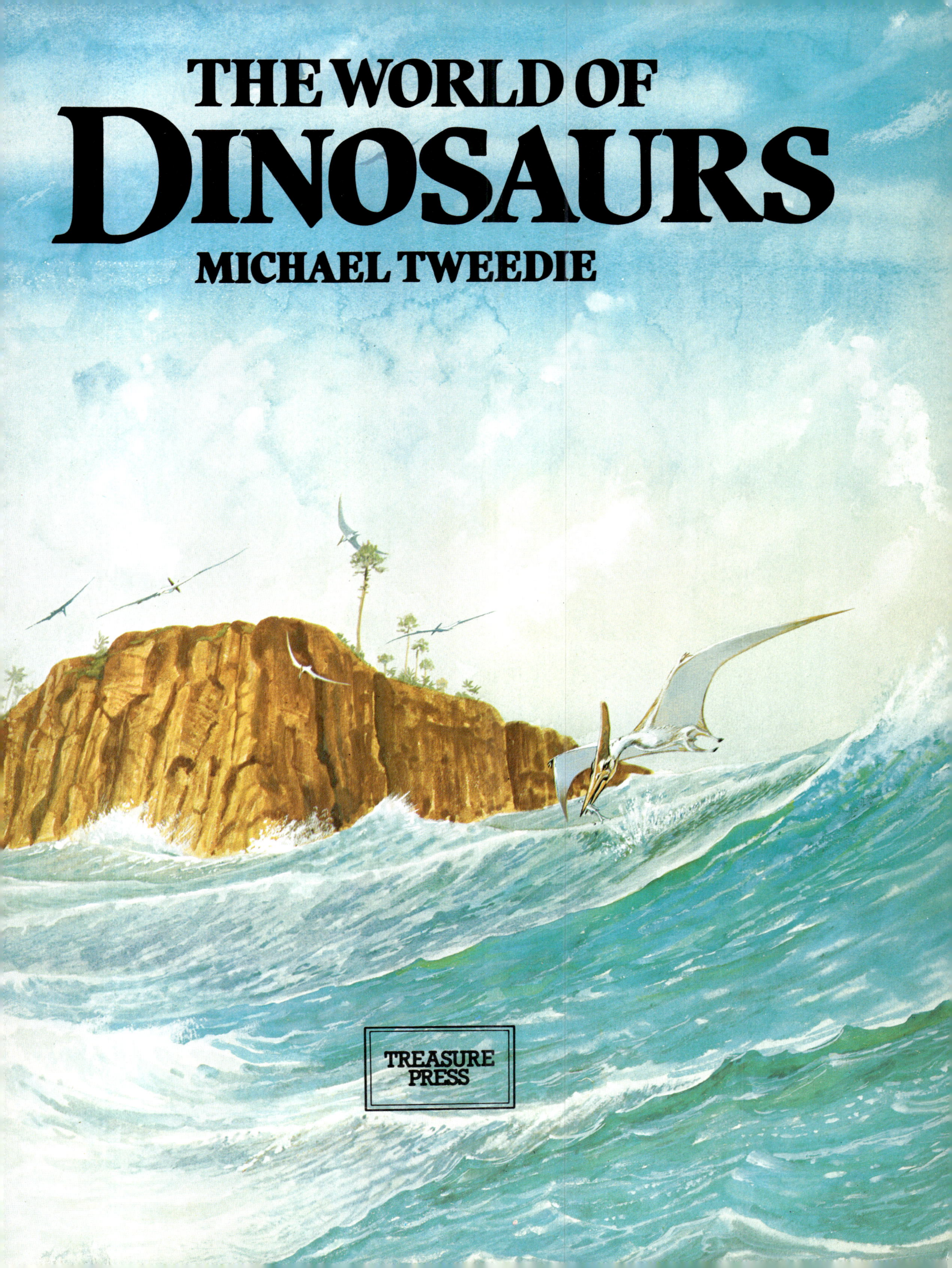

TREASURE PRESS

PREVIOUS PAGES
The great pterosaur, *Pteranodon*, probably had a way of life similar to that of the modern albatrosses, soaring effortlessly over the ocean, feeding on fish and breeding on small islands.

First published in Great Britain by Weidenfeld and Nicolson Ltd

This edition published by Treasure Press
59 Grosvenor Street
London W1

Reprinted 1983

ISBN 0 907407 63 3

Printed in Hong Kong

Contents

Chapter One

Background to the Dinosaurs

Plants and animals have inhabited the surface of the earth for an immensely long period, within which man is a newcomer. His appearance in his present form took place a mere yesterday ago compared with the times of the creatures with which this book is concerned. We cannot literally look back in time to see what they were like and how they lived, but the study of geology reveals, by indirect means, a remarkably vivid picture of them.

Geology is concerned with the rocks which make up the outermost layer or 'crust' of the earth, and with all the evidence which they contain of the earth's history. They are classified broadly into three kinds. Igneous rocks are those which have solidified from the molten state, either deep down in the crust (granite, gabbro) or at the surface as the result of outpouring of lava from volcanoes. Sedimentary rocks are formed by the accumulation of coarser or finer particles worn or eroded from the land and usually carried by rivers to the sea, where the material sinks to the bottom. Finally, the metamorphic rocks are those which result from alteration of either of the two primary types by heat, derived from igneous activity, or by the enormous pressures associated with earth-movement and mountain-building. Slate, schist and gneiss are metamorphic rocks.

In this book we are concerned only with sedimentary rocks, and we need to understand the processes of sedimentation which operate all over and around the continents, because the preservation of fossils is associated almost exclusively with this branch of geology.

The turbid waters of a river bear obvious witness that the substance of the land is constantly being eroded away and carried down to the sea, where it sinks to the bottom to form layers of sand, silt or mud according to the size of the mineral grains being deposited. Much of the finest sediment is carried far out and deposited on the ocean floor beyond the continental shelves. Coarser sediments, and sometimes mud as well, accumulate on the continental shelves: beach nearest the shore, sand further out and mud in deeper water lacking strong currents. In seas where there is a minimum of detritus (suspended rock particles) carried out from the land, calcareous or chalky sediment will slowly accumulate, partly by chemical precipitation, partly consisting of innumerable tiny shells of various organisms. The globigerina ooze of the present day illustrates the formation of such a sediment.

Often areas of the continental shelf on which sedimentation is taking place slowly sink, due to earth-movements, or simply to the weight of the accumulating sediment, and great thicknesses of it may be formed without much change in the depth of the water. As this process goes on the deeper layers will be compressed by the weight of the overlying deposit and will turn to stratified rock, sandstone, clay or shale; calcareous sediments result in the formation of limestone or chalk. Sandy and shaly deposits also accumulate in large freshwater lakes and on flat plains over which rivers pursue a meandering course. Windblown sand may also accumulate in desert areas. In every case slow subsidence can lead to great thicknesses of strata being formed.

The Study of Fossils

The seas, fresh waters and lands of past ages were inhabited by animals and plants, as they are today. Animals that lived in the water, or got into it accidentally, would sink to the bottom and their shells, bones or teeth would become embedded in the accumulating sediment. After uplift of the strata these can be found as fossils where the rock is exposed in cliffs or river-beds or by artificial excavation. Most animal fossils represent the hard parts, shell, bone or teeth. These are seldom preserved in their original form, but only after they have become impregnated by minerals carried in solution by percolating water. In this way they are petrified and their chemical composition is completely altered, but their shape and often their internal structure is preserved. Shales are often compacted by the weight of overlying strata, and fossil bones in them will be compressed; whole skeletons of marine reptiles are found flattened in this way.

The branch of geology that is concerned with fossils is called palaeontology. Evolution has produced a succession of forms of life on the earth, the great majority of which have become extinct at various stages in geological time. The earliest forms were very small, simple and confined to the sea, the later of increasing diversity in all respects including size; some of these left the sea and became inhabitants of the land and inland waters. Sedimentary rocks of all ages are exposed at the surface of the earth, and they are found to contain fossils which date up to the recent period from the time, hundreds of millions of years ago, when animals developed hard parts suitable for preservation. In rocks of earlier date than this fossils are extremely rare. Since sedimentation proceeds from below upwards the older sediments will in almost all cases underlie the later; and since the general course of evolution has been the same all over the earth, the fossiliferous rocks can be dated, relatively to each other, by identifying the fossils they contain. An enormous amount of work based on these two principles has established a world-wide geological record or fossil record, which is summarized on page 10. The fossil record is divided into three eras, the Palaeozoic, Mesozoic and Cenozoic, meaning eras of ancient, middle and modern kinds of life. Each era is divided into periods. The record is shown with the first or oldest part at the bottom of the page because this is the way in which the stratified rocks, on which it is based, are disposed.

As the time chart on page 10 shows, life was at first confined to the sea. The plants were algae or seaweeds, the animals all of extinct species, though some of them were not unlike their modern counterparts. Molluscs, echinoderms and brachiopods still exist, the two first in abundance; trilobites are no longer alive, but they were certainly arthropods as modern crustaceans, arachnids and insects are. Graptolites, appearing usually as mere scribble-marks on the rocks, are mysterious creatures with no close relatives alive today, though they probably lie near the same line of evolution as the vertebrates or backboned animals. Plants probably moved out on to the land in the Ordovician period, and were established there in the Silurian. They were soon followed by invertebrate animals such as scorpions and insects. The first vertebrate animals appear as fossils near the end of the Ordovician; these were jawless fishes related to the modern lampreys. Fishes with upper and lower jaws followed and flourished in the seas and rivers in Silurian and Devonian times. One group of them, the lobe-finned fishes, gave rise by evolution near the end of the Devonian period to the earliest amphibians, which were the first vertebrates to have the land as their normal habitat, though they could only breathe air after a juvenile period passed in the water; this way of life has persisted in the modern amphibians, whose young are well known as tadpoles.

The passage from amphibian to reptile came in the Carboniferous, and for the first time there were wholly terrestrial vertebrates, eggs, young and adults alike. From the Permian to the end of the Cretaceous, reptiles (or animals generally regarded as reptiles) were masters of the dry land, and also assumed something like mastery of the surface waters of the sea. At first the group collectively known as the 'mammal-like reptiles' prevailed, the pelycosaurs in the Permian and their more advanced descendants the therapsids in the Triassic; during their decline in the late Triassic these gave rise to the true mammals. But during this period both the therapsids and their descendants were overtaken and eclipsed by a far more active and efficient type of reptile which was ancestral to the great assemblage of archosaurs; these comprise the crocodiles, which remained wholly reptilian, and the pterosaurs (or pterodactyls), the dinosaurs and the birds, which did not. The earliest birds appeared long after the mammals, in the late Jurassic.

The evolutionary history of the main groups of reptiles, from the Permian period to the present day, is summarized in the diagram on page 17. This and the time chart on page 10 should be used for reference in reading the chapters that follow this one.

How Fossils are Formed

By far the commonest fossils are provided by the shells and other hard parts of marine animals that lived in seas where cumulative sedimentation was taking place. Many limestones consist largely of molluscan shells or remains of corals, which can be beautifully displayed by grinding and polishing flat surfaces of the rock. Clay and shale of marine origin almost always contain minute fossils of foraminifera, the shells of protozoan, or single-celled, animals. These can be revealed by carefully washing a handful of the rock (geologists regard them as rocks) until all the mud is removed and only a fine gritty residue remains from which the fossils can be picked out under a microscope. In most types of marine strata isolated teeth and bones of vertebrates may be found, but they are generally rare. The branch of palaeontology specially concerned with the remains of large vertebrate animals, such as dinosaurs, is a highly specialized one. The best approach to it is to consider how creatures of this kind come to be preserved as fossils.

Most of the dinosaurs about which we have any information were inhabitants of the forested flood-plains of rivers; some probably lived in associated lakes and marshes or in the rivers themselves. The carcass of a dinosaur destined to provide a fossil had first to get into a river; those that died on dry land were quickly broken up and dispersed by scavengers. It might fall into the river accidentally, it might be dragged in by a gigantic crocodile, or it might be an aquatic species of dinosaur. The body would sink at first, but later float when gases of putrefaction accumulated inside it. If the great carcass was carried to the river-mouth and out to sea it would disintegrate slowly and fall piecemeal to the bottom, providing fossils consisting of single teeth and bones. It might sink or be stranded in the delta, whole or in pieces, and be

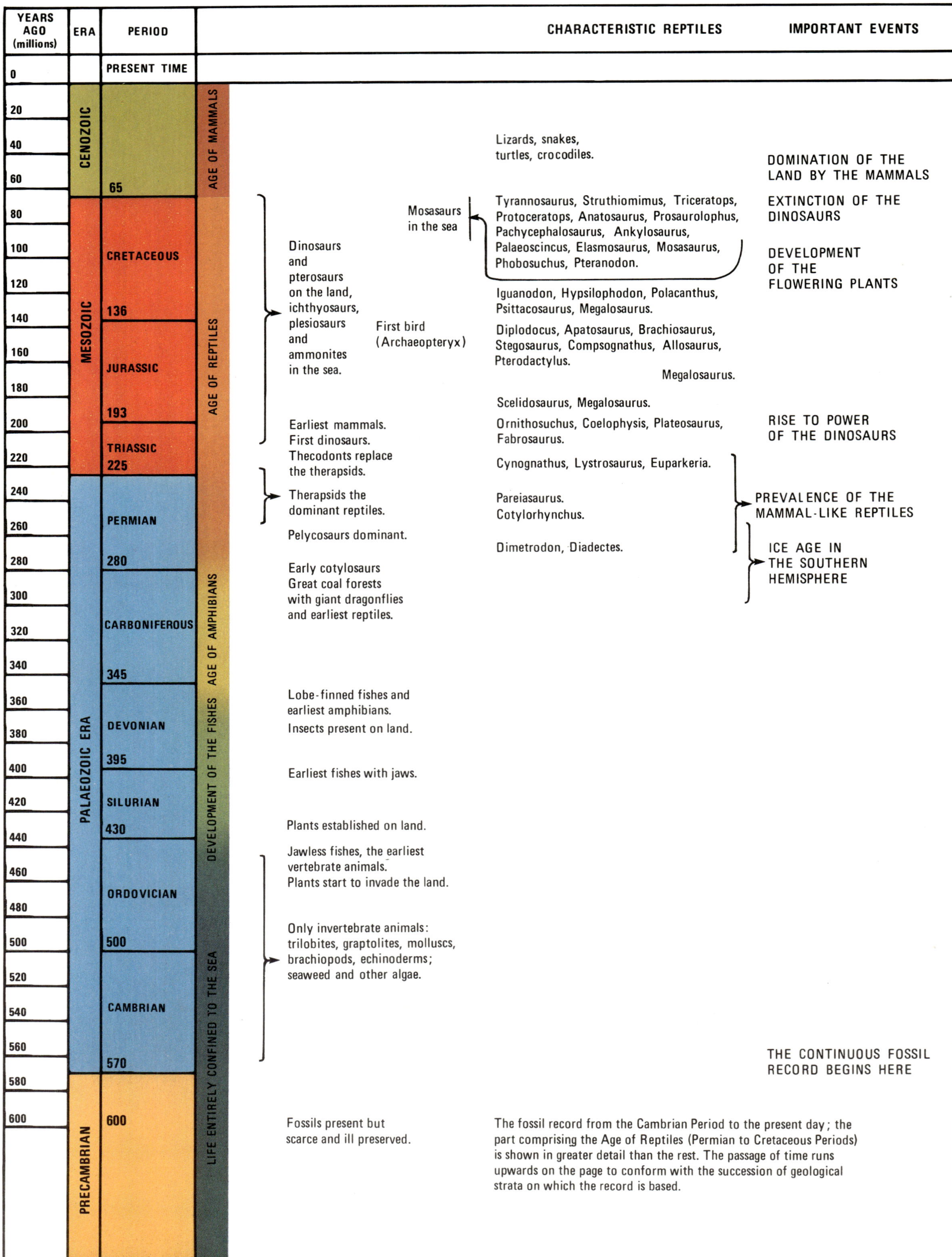

The fossil record from the Cambrian Period to the present day; the part comprising the Age of Reptiles (Permian to Cretaceous Periods) is shown in greater detail than the rest. The passage of time runs upwards on the page to conform with the succession of geological strata on which the record is based.

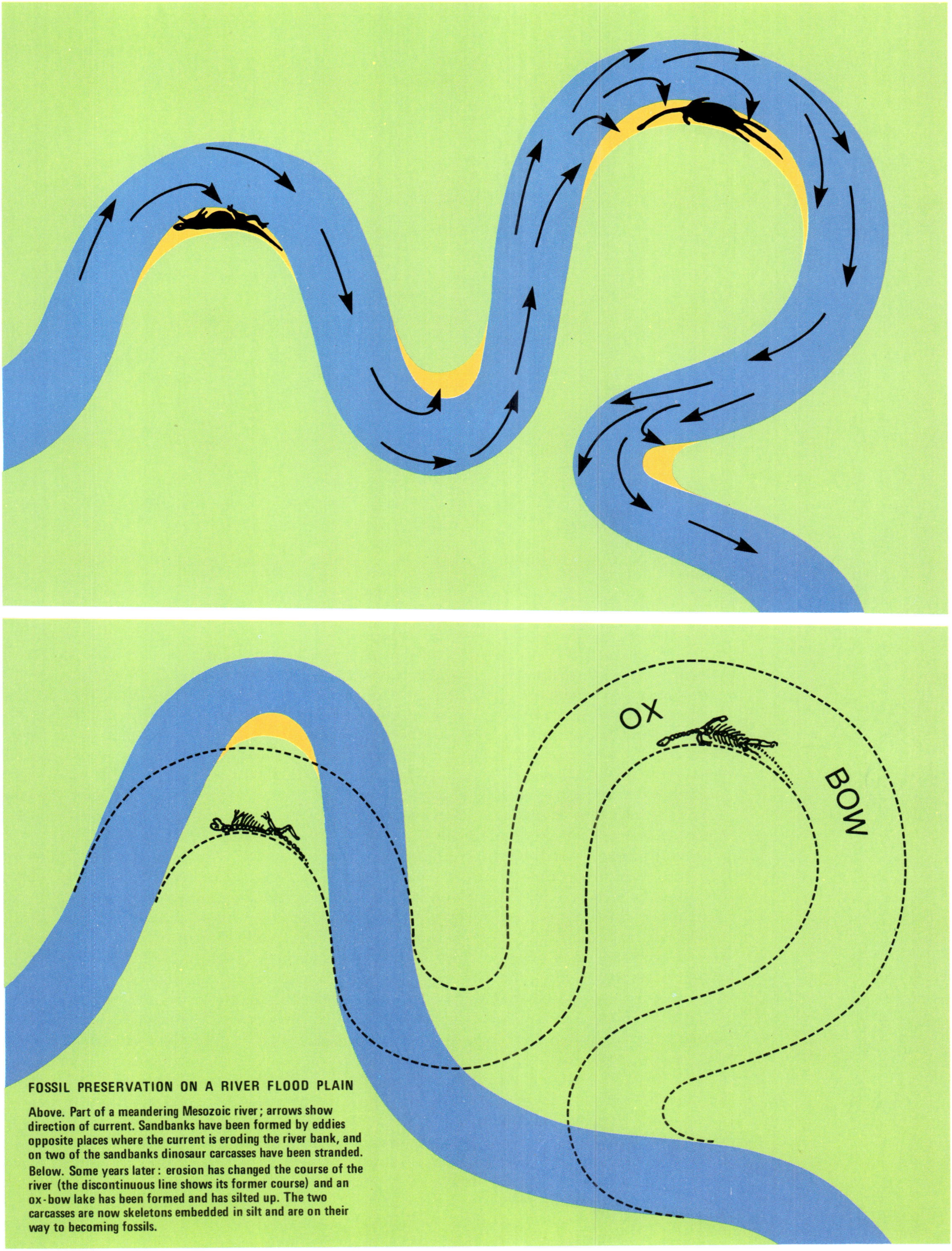

FOSSIL PRESERVATION ON A RIVER FLOOD PLAIN

Above. Part of a meandering Mesozoic river; arrows show direction of current. Sandbanks have been formed by eddies opposite places where the current is eroding the river bank, and on two of the sandbanks dinosaur carcasses have been stranded.

Below. Some years later: erosion has changed the course of the river (the discontinuous line shows its former course) and an ox-bow lake has been formed and has silted up. The two carcasses are now skeletons embedded in silt and are on their way to becoming fossils.

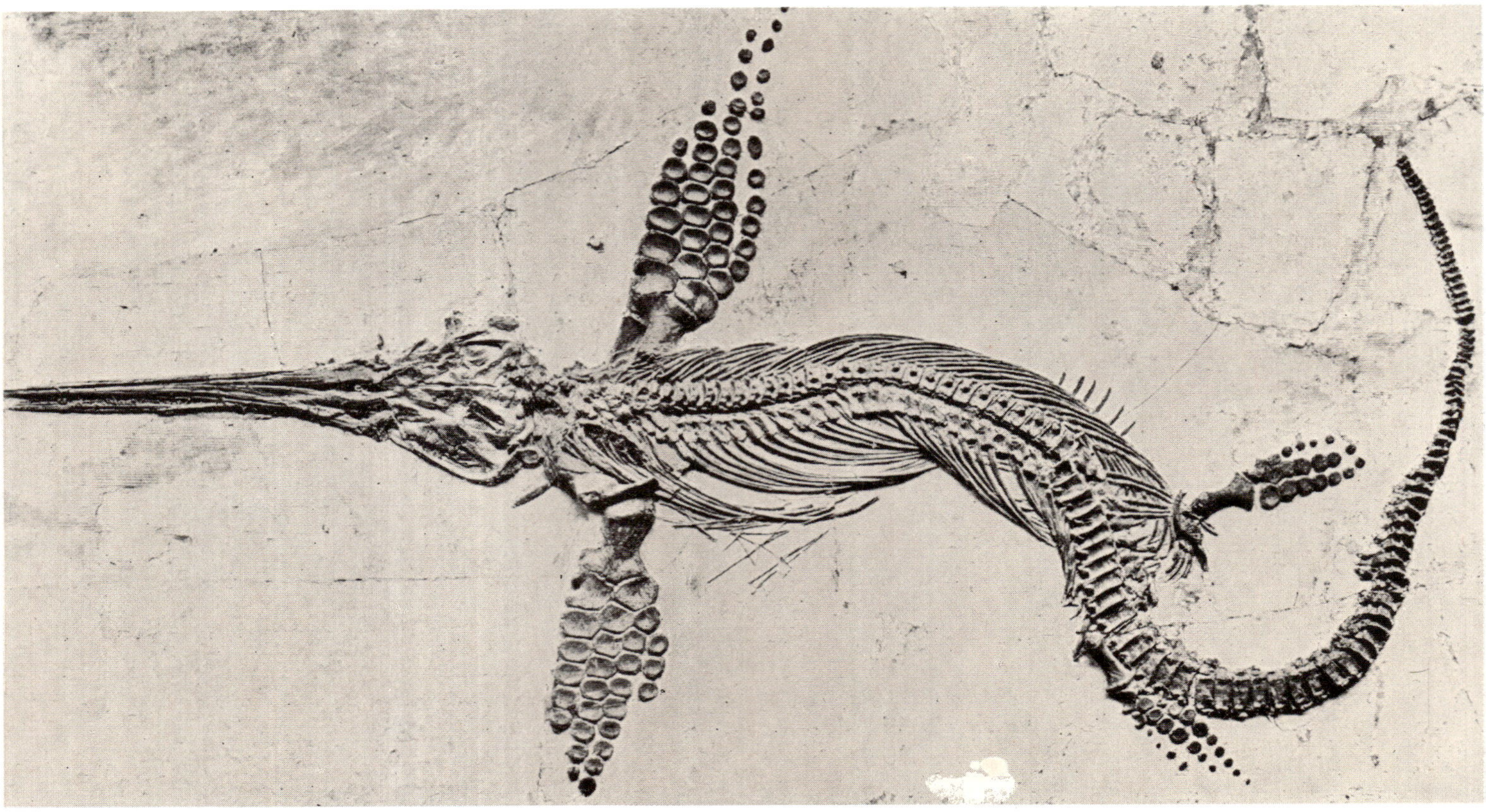

This ichthyosaur fossil is an example of the wonderful perfection with which marine reptiles are sometimes preserved.

covered up by the formation of a sand-bank. Or it might come to be entombed in the way explained by the diagram on page 11.

On river flood-plains a special type of sedimentation takes place. A river running across a level plain never pursues a straight course. This is because any chance deviation from straightness is enhanced by the action of the current, which scours the concave bank, while eddies in the water lead to deposition on the opposite side. This leads to the exaggerated loops called meanders, to ox-bow lakes, and to the proposition that a river occupies, at one time or another, the whole area of its flood-plain. The diagrams show this better than words, and they also show how carcasses of large animals floating in the river can be stranded at places where silting is taking place. If the area is undergoing subsidence the river will flood over the whole plain in times of heavy rainfall, depositing sand and silt, and the embedded skeletons will be more and more deeply buried in sediment and will in due course be fossilized.

Other hazards that have given us dinosaur fossils occurred when the creatures fell to their deaths in ravines or died in deserts and became covered with silt deposited during a sudden flood or by wind-blown sand. These bodies are sometimes mummified and details of the skin preserved. Some of the most perfect vertebrate fossils are those of large marine reptiles. Preservation of these depended on the body reaching the sea-bottom intact at a depth where low temperature and high water pressure would prevent them from floating due to putrefaction. The other necessary factor is of course rapid sedimentation to bury the carcass.

It might be supposed that the preservation of dinosaur footprints would be a rare freak of fossilization, but fossil tracks of them are in fact not uncommon. The first traces of dinosaurs to be noticed were tracks on rock pavements of Triassic age in the Connecticut valley, eastern USA. As early as 1800 they were well known to the local people and given a biblical attribution, 'Noah's raven'. They are in fact mostly large three-toed prints and were regarded as bird tracks by the first serious student of them. Rather similar footprints of *Iguanodon* are found in the lower Cretaceous Wealden strata of south-eastern England. The most dramatic dinosaur track was found in Texas and is that of a 'brontosaurus' type of dinosaur, each footprint 3 feet (1 m) long and with a capacity of 81 litres. These footprints, especially if a continuous trackway can be found, tell us a great deal about how the dinosaurs walked, their length of stride, how their bodies were poised over their legs and whether they went on four legs or two. Of course they cannot always be exactly correlated with the skeletal fossil remains (fossil footprints and bones are not often found near each other) but approximate identification is possible.

Their mode of origin and preservation can best be explained by reconstructing a suitable set of circumstances. Imagine a large lake in the Mesozoic lowlands, whose area of water changes with the seasons. Early in the dry season a wide margin of mud already surrounds the lake. A dinosaur goes over the mud to drink, leaving deeply impressed tracks, and later the imprinted mud dries and hardens under the hot sun, the more so if it is limey and slightly sandy. During the dry season the mud will harden quite deeply in and around the tracks. When the wet-season floods set in the water will rise and cover the tracks and they may be filled and overlain by muddy sediment. For preservation the normal concomitant of sedimentation is needed, slow sinking of the land surface, so that the lake extends and perhaps eventually connects with the sea, and many feet of sediment accumulate above the level of the footprints. If the strata are later raised up and eroded, so that the stratum including the footprints is exposed in the bed of a river or on the floor of a quarry, the shale representing the mud that filled in the tracks will wash out of them, and there they will be, just as the dinosaur made them, but imprinted in hard rock. They are exciting fossils because they are the remains of a living animal, not a dead one.

Hunting for Fossils

Finding, extracting and studying fossil bones and skeletons is a highly specialized science. Sometimes the bones are noticed and reported by miners or quarrymen before extensive damage has been done; in such a case any responsible manager will stop work immediately and call in professional palaeontologists who will see if the find is of value, and if it is will ask for permission and time to recover the bones. In 1878 a find was made in a Belgian coal-mine of *Iguanodon* skeletons, and three years were spent in recovering them, throughout which the mining authority cooperated fully with the museum at Brussels.

Most of the important dinosaur material has been found by exploration in the field. The hunter must of course go to an area where rocks of Mesozoic age are exposed at the surface. The best sort of country to work in is the kind known in America as 'badlands', hilly, rocky desert or semi-desert, where rock is exposed without much in the way of soil cover. The dinosaur-hunter must be a strong, active person capable of walking many miles a day, and must have a sharp eye and long practice in spotting fossil bone and distinguishing it from the surrounding rock. If fragments are found at the base of a cliff, the cliff face must be searched for the stratum from which they have weathered out. Experience in rock-climbing is part of his or her professional competence. When a find is made it is seldom possible to be sure whether it is of value. A vertebra protruding from a rock face may be a single bone from a carcass dispersed soon after the animal died. If it is part of a skeleton, close examination and some excavation is needed to tell whether the greater part of the skeleton or just the end of the tail lies behind it.

The methods used in recovering dinosaur remains bear no resemblance to casual fossil-hunting, and involve principles very similar to those of archaeological excavation. First as much as possible of the skeleton is exposed by careful removal of the rock embedding it. By measurement, drawing and photography the exact position of every bone is recorded. The bones are nearly always brittle and any attempt to dig or prise them out will shatter them to a degree which makes reconstruction difficult or impossible. First they must be hardened by painting or spraying on dissolved resin of some kind; shellac was used formerly but has been superseded by modern compounds such as alvar. Next wet tissue-paper is laid over all bone surfaces, and then pieces of hessian, dipped in a strong mixture of plaster of Paris, are applied overlapping. If the bone is a long one splints of wood are attached to it on the same principle as those used to set a living broken limb. When the top of the specimen is completely covered the rock around and under it is chiselled away until the bones rest on a small pinnacle. This is then broken and the whole block turned upside-down and given the same treatment underneath. When the plaster has set it is ready for transport. In badlands it is often impossible to bring vehicles close up to the excavation, so the blocks must be kept small enough to be carried on slings or stretchers. An entire dinosaur skeleton has to be divided into convenient blocks and taken out in pieces, an exacting task if the skeleton is closely articulated.

When the material arrives at the museum workshop the plaster bandages are first soaked and removed and the bones then freed from the enclosing rock, which may be a very delicate and difficult business. Small hammers and chisels are the basic tools, but special refinements have been invented. Among these is a tool which delivers vibrations against the rock at a rate of thirty thousand a second, a sort of ultrasonic chisel, and an air blast of abrasive powder cuts away any hard surrounding material quickly and safely. Chemical methods are increasingly used involving acids which will dissolve the enclosing rock but not the fossil bone. If the rock is

OVERLEAF
The forests from which most of the world's coal was formed existed about 300 million years ago. The dominant trees were gigantic horse-tails and club-mosses. Insects, including huge dragonflies, flew among them. Amphibians and the earliest reptiles lived in the swamps in which they stood.

The skeleton of *Tyrannosaurus rex* reconstructed from fossil material in North America. The skeleton is displayed in the animal's most probable standing posture against a background of Upper Cretaceous scenery.

limestone this is specially useful, as the calcium phosphate of the bones is much less soluble than calcium carbonate, of which limestone mainly consists.

Every well-preserved bone bears roughened scars where tendons were attached, articulating surfaces showing what sort of movements could be performed by the joints, and sometimes holes or foramina betraying the course of nerves and blood-vessels. By study of these, comparing one with another and with the corresponding bones of living animals, our knowledge of dinosaur natural history is built up. Teeth are particularly valuable as they always give information about the kinds of food their owners ate.

Specimens for display in museum galleries have to be complete, and often it is necessary to build up a skeleton from two or more specimens of the same species and of comparable size. Careful modelling of missing left- or right-hand bones from ones that are available is also done. Some very perfect specimens have had moulds made of each bone so that identical casts can be taken and replica skeletons prepared for other museums. Modern casting is done with glass fibre and resin rather than plaster. This is strong and light and requires a less robust and conspicuous metal framework to support the specimen. A skeleton may require several years of continuous work between arrival at the museum and the moment when it goes on exhibition.

Geological Time

Below the names of the periods on the time chart (page 10) figures are entered indicating the date of the beginning of each period in millions of years ago. Fifty years ago this could not have been done, though the geological record had already been established, and the periods named, for the greater part of a century. This record gave a fully recognizable moving picture of the course of evolution followed by animals and plants, but the methods which produced it attached no absolute time scale to the fossil record. Attempts were made to remedy this by measuring the total thickness of the strata (which varies from place to place anyway) and by calculating how long the sea should have taken to become as salt as it has done, but it was realized that such estimates could not be reliable. During the 1930s methods were developed of dating the geological record, based on the rate of disintegration or 'decay' of certain radioactive elements. This is known to be absolutely constant regardless of temperature or any other physical conditions, and in some cases it is exceedingly slow. The first elements to be used were uranium and thorium, and ordinary uranium (the isotope 238) has a 'half-life' of 4.5 thousand million years. This means that a crystal of a mineral containing uranium which formed that length of

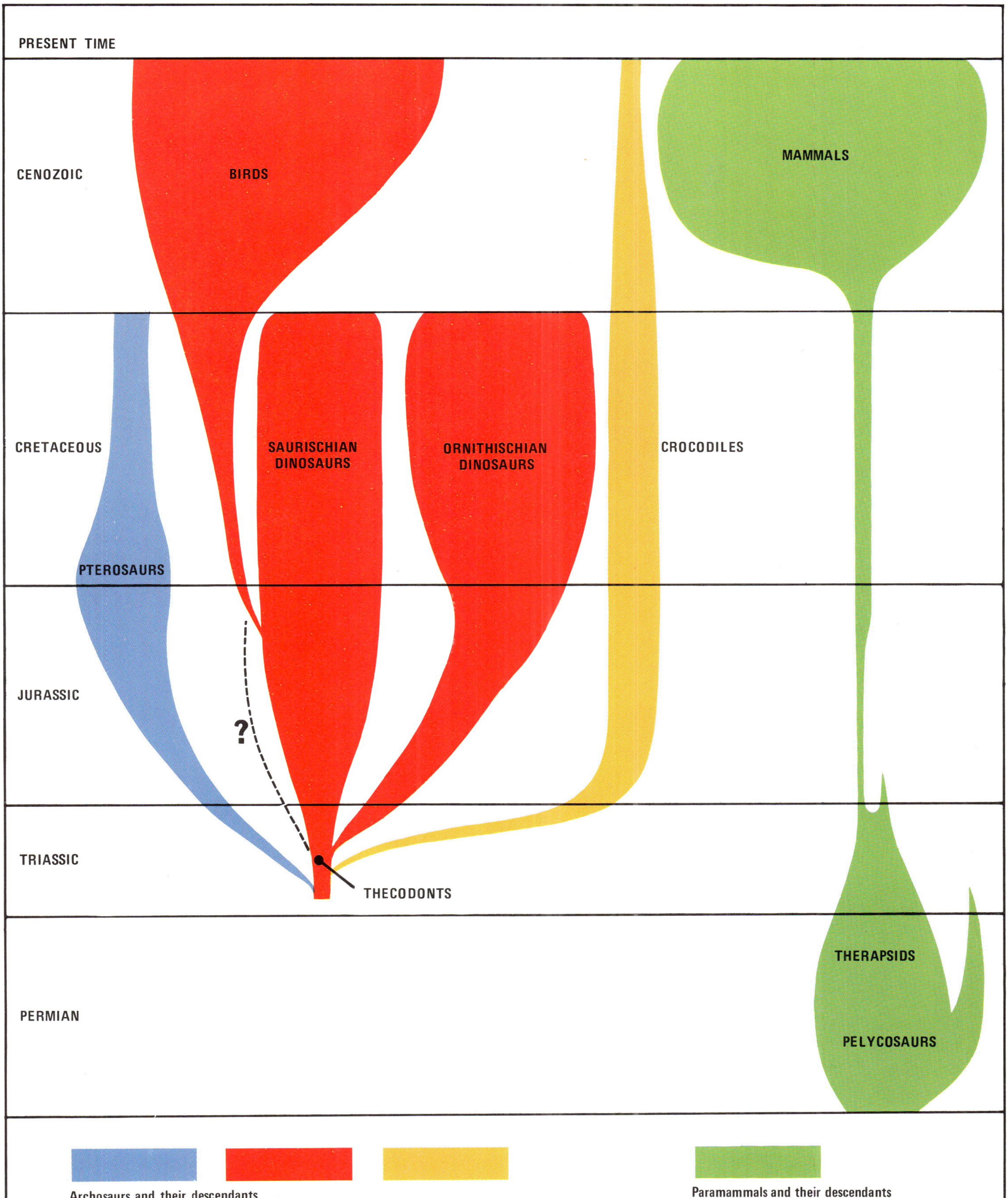

Diagram of the evolution of the more important groups of archosaurian and paramammalian animals. In this diagram preference is given to the theory of the origin of birds from saurischian dinosaurs.

The skeleton of a brontosaur being mounted at The American Museum of Natural History near the turn of the century.

time ago (i.e. about the date of the birth of the solar system) would now contain just half as much uranium as it started with; the rest would have been converted at a perfectly constant rate into helium and lead. In the next 4.5 thousand million years the amount would be halved again, so that a quarter of the original uranium would remain. Thorium also has a long half-life and also has lead as an end-product, and both uranium and thorium produce lead of particular isotopes, which can be identified. By ascertaining the amount of uranium and the amount of the relevant isotope of lead present in a sample of rock an estimate can be made of the date at which the rock crystallized.

The disadvantage of this method is that uranium and thorium minerals are rather rare, and that it can only be applied to igneous rocks. If these are intruded into fossil-bearing sediments, or are lavas overlying them, then obviously the sediments are older; if the sediments overlie the igneous rock they are younger. But how much older or younger it is often difficult to say, so uranium and thorium are not very suitable for dating the fossil record. The same objection applies to the breakdown of radioactive rubidium to form strontium.

In recent years these methods have been largely superseded by the potassium-argon method. This depends on the presence in normal potassium of a small but constant proportion of a radioactive isotope, potassium 40. It decays to form either normal calcium or argon 40, which can be distinguished from ordinary argon, and measurements of the amounts of potassium 40 and argon 40 in the rock sample are used to determine its age. Potassium is a common element in rock of all kinds, and the age of volcanic tuffs, or consolidated volcanic ash, can be determined; these are often interstratified with fossil-bearing sediments. The sediments themselves cannot be dated because the radioactive history of each crystal grain in a sediment starts with the cooling and solidifying of the igneous rock from which it was ultimately derived by erosion.

The earliest date entered on the time chart is 600 million years ago, rather before the beginning of the Cambrian period when the continuous fossil record begins. Some rather obscure fossils of animals are known from strata older than this, and there are fossil reefs made by green algae 2 thousand million years old. The very earliest signs of life, fossil cells of bacteria or blue-green algae, are 3 thousand million years old, and there is evidence from both the earth and the moon that the solar system came into being 4.6 thousand million years ago. Going forward from the Cambrian we find the first fishes at about 440 million years, the earliest reptiles around 300 million and the dinosaurs dominant from 200 to 65 million years ago.

We are accustomed to look at time in terms of human life-spans; a thousand years ago seems quite remote. Geological time can be dimly appreciated by correlating it with a scale that forms part of our experience. Let us consider time on a scale where one yard equals a thousand years, so that the time of Jesus Christ and Julius Caesar is six feet away. On this scale the beginning of the Cambrian period would be 320 *miles* away. The dinosaurs' reign would start 113 and end 37 miles away. The earliest signs of life would be distant 1700 miles, and the birth of this and the rest of the planets 2600 miles away.

Chapter Two

The Age of Reptiles

OVERLEAF

Dimetrodon was a reptile belonging to the group known as pelycosaurs. It had a 'sail' of skin supported by a row of bony spines growing along its back; this probably acted as a heat regulator both for warming up and cooling down the animal's body. *Diadectes*, which lived at the same time, was one of the primitive reptiles called cotylosaurs.

The vertebrate animals, including ourselves, are descended from fishes, and the story of the reptiles can be started conveniently from the time when the first fishes left the water and learned to breath air and walk on the land. During the Devonian period the lobe-finned fishes or rhipidistians lived in rivers and lakes which periodically dried up. Fishes isolated in a drying pool will die when the water becomes stagnant and deficient in oxygen. Some of them 'invented' a new way of breathing by taking air into an internal sac or 'lung', connected with the pharynx, as well as passing water over the gills. In modern fishes this sac has been turned into the swim-bladder, a hydrostatic organ enabling the fish to control its overall specific gravity. The lobe-finned fishes also had pectoral and pelvic fins reinforced with bone so that they could be used for pushing on land as well as for swimming. By this means they could leave a drying, shrinking pool in search of a better one, and they were the ancestors of the amphibians, the first land vertebrates. It must be realized that the fishes had no 'intention' of invading the land; all they wanted was to find a healthy pool where they could go on being fishes. Natural selection, the mainspring of evolution, directs itself only to immediate problems; sometimes by chance an evolutionary innovation will serve a purpose beyond the one for which it was first developed.

The amphibians, represented today by such animals as frogs and salamanders, developed rapidly during the Carboniferous period and some became as large as crocodiles. They were very abundant in the great swamps in which much of the world's coal was formed. The amphibians never went far from water; they had to lay their eggs in it and their young, like modern tadpoles, passed their early lives living and breathing in the water just as fishes do. The water at that time swarmed with predatory amphibians, all ready to eat each other's eggs and young, and with large fierce insect larvae as well. To make sure that some of their offspring survived they had to lay great numbers of eggs, which was biologically wasteful. To meet this situation it seems most probable that some amphibian evolved a type of egg that could be laid on land, hidden away and relatively safe from enemies. It would have to be larger and more complicated than the egg of a frog or a newt, and would also have to be fertilized inside the female's body, not after leaving her body as amphibians' eggs normally are.

There is no fossil evidence for this, nor ever likely to be, but we can assume that it happened because an egg conforming to this specification is laid by reptiles and by birds as well. It is called the amniote egg and is worth close examination as it is the symbol of an evolutionary breakthrough of the first importance. It is covered with a shell or parchment-like skin, slightly porous but adequate to prevent drying up if the egg is laid in humid surroundings. Inside are three membranous bags, the amnion containing liquid in which the embryo develops, the yolk-sac containing a store of concentrated food, and the allantois which serves as a lung in contact with air that diffuses through the shell. When the shell is calcareous, calcium carbonate is dissolved out of it and incorporated in the growing skeleton of the embryo. The young animal grows up in a little private pond hidden away and well-provisioned and aerated. When it hatches it is far past the gill-breathing tadpole stage and is a small replica of its parents, well able to live on land and fend for itself.

The oldest fossil reptiles, which resemble amphibians in many ways but are distinguished by skeletal characters, date from the middle Carboniferous, while the earliest fossil reptilian egg is from the Lower Permian, a good deal later. Eggs, however, are not often preserved as fossils, and the amniote egg is such an essential feature of starting to be a reptile that it is reasonable to suppose that development of it was the first step in that direction. This progression was, of course, by far its most important consequence, but remember that it began as an adaptation directed towards being a more efficient amphibian, not towards turning into a reptile.

Just which of the Carboniferous amphibians gave rise to the reptiles is uncertain, though a creature called *Seymouria* is often cited as a transitional form between the two Classes. It lived, however, in the Permian, much too late to be an actual ancestor of the reptiles, and recent research rules it out on other grounds as well.

The First Reptiles

The earliest reptiles belonged to a group called cotylosaurs which persisted into the Triassic. A few attained a fair size; the late Permian *Pareiasaurus* was 7 feet (2 m) long and heavily built, but the cotylosaurs themselves never became dominant animals. They are believed, however, to be the ancestors of most if not all of the other reptile groups. One of these, the pelycosaurs, became very abundant and diverse early in the Permian, and with them the age of reptiles started. This expression has generally been regarded as applying to the Mesozoic era, but reptiles assumed complete dominance of the land in Permian times, and it is much more realistic to include this period as the first stage in the reptiles' 200-million-year reign over the lands and surface waters of the earth.

The best-known and one of the largest of the pelycosaurs is *Dimetrodon*, whose fossil remains are found in the Lower Permian red beds of Texas. It was a carnivore that grew to 10 feet (3 m) in length, and the strata in which it is found suggest that it lived in a dry climate approaching desert conditions. Most reptiles have the teeth all of one type, but *Dimetrodon* possessed the three types characteristic of mammals, canines for killing its prey, incisors for biting off pieces of meat, and cheek-teeth for cutting up or mincing the meat before swallowing. The strangest feature of *Dimetrodon* is the presence of long spines of bone projecting upwards from its vertebrae, longest over the middle of the back, grading down towards head and tail. These spines must have supported a web of skin like a sail, and in a desert environment this could have served the animal in the following way.

Dimetrodon was certainly a cold-blooded or ectothermic animal whose temperature, like that of lizards and snakes, varied with that of its surroundings. The effect of a lowering of temperature is to make such animals sluggish; at sunrise, after a cold desert night, a dimetrodon would be incapable of active movement and, being large, would take a long time to warm up in the sun. It is believed that the 'sail' was permeated with blood-vessels and acted as a solar heater. If the animal turned sideways to the rising sun and dilated the blood-vessels leading to and from the sail its body would warm up much faster than if the sail were not there, and it could start pursuing animals less well equipped and still inactive. In the heat of the day it could turn head or tail to the sun and use the sail to radiate away excess heat. At night contraction of the blood-vessels would minimize heat loss from the sail. Small reptiles living in the desert burrow in the ground to escape extremes of heat and cold, but *Dimetrodon* was much too large to do this.

There were large herbivorous pelycosaurs as well in the Permian. *Cotylorhynchus* was nearly 10 feet (3 m) long and very massive. Its bulk must have helped to keep its temperature constant; a large body has a small surface area relative to its mass, so heat passes in and out slowly.

The pelycosaurs were supreme for the first half of the Permian, 30 to 40 million years, but were then replaced by a group of their own descendants, the therapsids. We noted the presence of mammal-like teeth in *Dimetrodon*, and in the therapsids various other mammal characteristics developed, and a gradation from them to the true mammals, which appeared near the end of the Triassic, can be clearly seen in the very rich fossil record that is available over this period of time. The pelycosaurs and the therapsids are known collectively as the mammal-like reptiles or paramammals.

There were two main groups of therapsids. The most abundant were the herbivorous dicynodonts which were the first successful group of plant-eaters among the vertebrates. Many of them had no teeth at all, some retained a pair of tusk-like upper canines, and they must have had horny edges to their jaws like those of modern tortoises. *Lystrosaurus* is a good example of a dicynodont; its remains are found in quantity in Lower Triassic strata in South Africa, India and China, and it has been discovered in Antarctica, giving support to the theory that these land masses were once joined and have separated by continental drift. It was 2 to 4 feet (about 1 m) long and had the tusk-like teeth in the upper jaw and the nostrils high up on the snout. It probably lived like a small reptilian hippopotamus, swimming and wading in swamps and breathing through its elevated nostrils.

Cynognathus, also Lower Triassic, is a good example of the other important group of therapsid reptiles, the cynodonts, which were carnivores. It was as large as a wolf, with dagger-like canine teeth, and the other teeth differentiated very much in the mammal pattern. The advanced cynodonts, such as this one, were the stock from which the mammals evolved, and they developed many mammal-like features. Their posture changed from the reptile pattern so that the legs were under the body and not sprawled out at the sides, and this greatly improved their ability to run. A secondary palate was developed in the roof of the mouth; this enables an animal to continue breathing while it is eating, and so makes it possible to chew up the food to prepare it for digestion. This in turn helps the animal to maintain a high level of activity. In reptiles the lower jaw consists of several bones. With evolution towards the mammals the dentary, in which most of the teeth are implanted, enlarges and the bones below and behind it become smaller and move to the back of the jaw. In mammals two of these bones, the quadrate and the articular, have left the jaw and turned into the incus and malleus, two of the three ear ossicles. These serve to convey and amplify vibrations from the eardrum to the internal ear. Reptiles have only one ear ossicle, the stapes; mammals inherit this and have added the other two to make up a very delicate and sensitive system. The possession of a lower jaw consisting solely of the dentary is regarded as a convenient arbitrary character on which to decide whether a fossil is to be called a

OVERLEAF
Lystrosaurus was an abundant and widespread animal that lived in swamps in early Triassic times. It was one of the therapsids or mammal-like reptiles and had a pair of conspicuous tusks in its upper jaws. The large thecodont *Erythrosuchus* (*right*) lived at the same time.

primitive mammal rather than an advanced therapsid.

It is probable that some at any rate of the cynodonts were covered with a coat of hair and therefore presumably warm-blooded or endothermic. Pits have been found on the bones of the snout, indicating nerve-endings leading to sensitive whiskers, and whiskers are modified hairs. *Thrinaxodon* was a small cynodont, well along the road to being a mammal, and an adult fossil specimen was found close to a young individual, suggesting maternal care and perhaps its concomitant, suckling with milk. If *Thrinaxodon* did this it would be a mammal by the definition of the living animal: 'a warm-blooded vertebrate that suckles its young', even though its lower jaw is reptilian by the palaeontological definition. The fact is that we cannot draw a line between the advanced therapsids and their mammalian descendants. Evolution is a continuous process and admits of no absolute boundaries or categories. It is only in cases where our knowledge of an evolutionary sequence is meagre and discontinuous that they seem to exist.

True and indubitable mammals appeared near the end of the Triassic. An exciting find of a complete skeleton of one was recently made in Lesotho. Named *Megazostrodon rudneri*, it was a little beast 4 inches (10 cm) long that must have looked rather like a shrew.

While the carnivorous cynodonts flourished in the early and middle Triassic the dicynodonts declined and an offshoot from the cynodonts, called gomphodonts, took their place as the dominant herbivores. At this time, then, the cynodonts and their derivatives, the gomphodonts and mammals, seemed poised for permanent domination of the land. But something went wrong: both cynodonts and gomphodonts declined and disappeared, leaving only a few trickles of descendants to carry on into the next period, the Jurassic. It is true that one of these 'trickles', the mammalian one, did eventually go on to dominate the land and to be the ancestral line leading to tigers and buffaloes and mice and men. But before making any conspicuous progress the mammals had to serve a humiliating 'apprenticeship' lasting 120 million years. No known Mesozoic mammal is larger than a cat, and most of the animals that maintained the mammalian line of descent through the long ages of the Jurassic and Cretaceous were rather like the modern insectivores: small, probably nocturnal creatures, burrowing, scuttling among rocks or climbing trees, rarely preserved as fossils, having one main preoccupation, to keep out of the way of dinosaurs.

The Coming of the Archosaurs

In the early Triassic, when *Lystrosaurus* was still floundering in the swamps and trying to avoid being caught and devoured by *Cynognathus*, a different type of reptile first appeared, quite unrelated to the paramammals. Because their teeth were implanted in sockets these reptiles are called thecodonts. The earliest of them were unremarkable, rather crocodile-like aquatic animals. There were some large ungainly land-dwellers among them, notably *Erythrosuchus*, whose skull was a yard long, making it the biggest known land animal of that time or previously. Some of the lake- and swamp-dwelling thecodonts evolved without much change into crocodiles. Others, possibly because they adopted a frog-like mode of swimming, developed long and strong hind legs, and some of these came out of the water and ran about on dry land. This was the first stroke of the bell that tolled the doom of the therapsids and drove the mammals into exile on the fringes of the Mesozoic environment.

In South Africa, still in Lower Triassic strata, fossils are found of a thecodont about three feet (1 m) long that has been called *Euparkeria*. It probably walked on four legs, but when in a hurry got up on its long, strong hind legs and ran. *Ornithosuchus* from the Upper Triassic of Scotland was rather larger and more advanced in this bipedal mode of running. Its fore legs were too small to be of use in locomotion and the structure of the pelvic girdle shows that the thighs were nearly vertical, bringing the legs under the body and giving a forward-and-back stride without any sideways sprawl. *Ornithosuchus* is usually classed as a thecodont, but at least one authority regards it as a primitive dinosaur. It does not matter at all – the important thing is that it gives us a good notion of how the dinosaurs came into existence.

It is true that today the fastest runners of all are quadrupeds, the cheetah and some of the antelopes. Mammals have a flexible backbone that will bend and straighten vertically like a spring, enabling the repetitive leaping gait known as a gallop. No reptiles ever acquired this, and their fastest quadruped gait is a rapid shambling or lumbering movement. But the ostrich is also among the fastest of running animals, and it runs not by flexing and extending its body in a gallop but by taking long springy paces with the two legs which are all it has. This is what the early dinosaurs did, and this ability to run rings round any therapsid was a major factor in their rapid rise to supremacy, which was already complete by Upper Triassic times, before the mammals had had time to radiate out from their small beginnings. More about this in the next chapter.

The thecodonts were the ancestors not only of the crocodiles and of the two great orders of dinosaurs, but of the pterosaurs (a more correct collective term than pterodactyls) and of the birds. Birds are so closely involved in the history of the dinosaurs that I will deal with their origin in a later chapter. The thecodonts and all their varied descendants are known collectively under the term archosaurs or ruling reptiles.

The only extinct archosaurs that remain to be discussed in this chapter are the pterosaurs, masters of the air until they were superseded by the birds. A fossil found in the late Triassic of Kirgizstan, north of the Himalayas, gives a clue to the origin of these wonderful animals. *Podopteryx mirabilis* was a small thecodont that had taken to the trees and become a glider, like the flying squirrels and lizards of today. This is a valuable asset to an arboreal animal for two reasons: it can leap safely into the air when pursued by a predatory climber, and it need never come to the ground. In continuous forest a gliding animal can go from tree to tree for an indefinite distance. In *Podopteryx* a web of skin stretched between the fore and hind limbs and between the hind limbs and the base of the tail. The latter web was the larger, but as evolution towards the pterosaurs proceeded, the one at the side of the body increased in area and acquired a leading edge consisting of the arm and the greatly elongated fourth finger of the hand. The first three fingers persisted as small clawed digits halfway along the wing, rather as the thumb does in bats.

Podopteryx was a sort of late Triassic living fossil, because when it existed pterosaurs had already evolved from earlier podopteryx-like ancestors. Many well-preserved fossils of them have been found in Jurassic and Cretaceous strata, and we can learn from them a great deal about what pterosaurs were like. They were not ancestors of birds: the two groups are independently evolved offshoots from the archosaur stem. Nevertheless, apart from the way in which their wings were formed, skin membranes instead of feathers, pterosaurs were remarkably like birds. This is because the problem the two groups had to solve, that of flying successfully, was identical.

Both birds and pterosaurs have hollow bones, and in birds this condition serves two purposes: it lightens the skeleton and forms a part of the very efficient breathing system that birds possess. Air is not merely drawn into and out of the lungs, it passes through the lungs into a number of air-sacs which lie inside the body and extend into the cavities of the hollow bones. In pterosaurs there are openings into the bones which strongly suggest that they had a similar breathing system. In typical reptiles the brain is small and does not completely fill the cavity in which it lies in the skull, so that an internal cast of the cavity does not represent either the full size or the exact shape of the brain. In birds and mammals the brain has enlarged and completely fills its cavity in the cranium, which in turn has enlarged to accommodate it. Some very perfect fossil pterosaur skulls show clearly that the brain completely filled the cranial cavity, and furthermore that the enlargement of the cerebral hemispheres, the cerebellum and the optic lobes had taken the same form as in birds. In both, the olfactory bulb is much reduced. Like birds pterosaurs required perfect balance and control of their movements and came to rely on their sense of sight at the expense of that of smell. It is safe to assume that they possessed an 'intelligence' far above that which we associate with any living reptile.

If, as these features suggest, pterosaurs had a metabolism comparable with that of birds they must surely have been endothermic or warm-blooded, a condition that, at any rate in small animals, requires an insulating cover such as feathers or fur. A claim was made long ago that signs of hair were apparent in some pterosaur fossils, and a find made in 1970 in Russia of a small Jurassic species settled the matter. The fossil is preserved in extraordinary detail, with impressions not only of the wing-membranes but of a thick coat of fur on the body and extending over the wings. Its discoverer named it *Sordes pilosus*. The fact is that pterosaurs were no more reptiles than birds are. If they were alive today we should give them Class status among the vertebrates equivalent to that of birds and mammals.

We have no direct evidence of their breeding habits, but no modern flying vertebrates, that is birds and bats, can fly in their infancy and practically all of them need parental care. It is difficult to envisage a hatchling pterosaur that could lead any sort of independent life, and quite reasonable to suppose that they resembled birds in this respect as well.

The earliest good fossils of pterosaurs are Lower Jurassic. These animals had fairly heavy skulls with numerous teeth and long bony tails, sometimes with a small rudder at the tip. The reduction and loss of both teeth and tail in later Jurassic and Cretaceous forms is a sign of the need to reduce weight to achieve efficient flying, the more necessary as they increased in size.

OVERLEAF

Left *Cynognathus* lived in the Triassic Period and was one of the therapsid reptiles which were the ancestors of the mammals. It was a carnivore, about as big as a wolf, and very likely had a coating of hair as one of its mammal-like features.

Right *Bienotherium* one of the late mammal-like reptiles, had reached a stage only a little short of being a mammal. It was a small animal and may have had a way of life rather like that of the modern rodents.

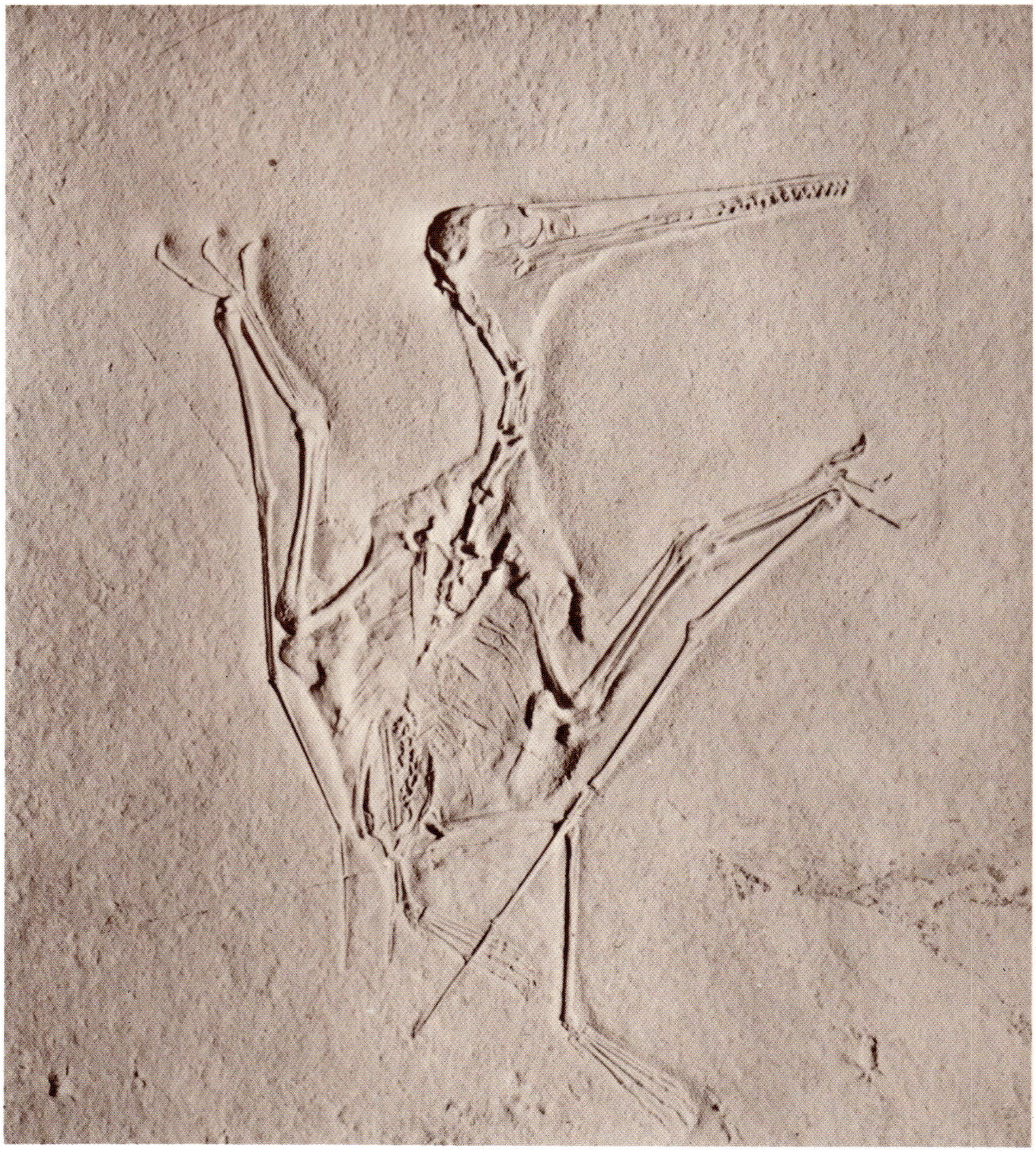

A beautifully preserved fossil of a small pterosaur (*Pterodactylus*) from the Lithographic Limestone of Bavaria. Impressions of the wing-membranes can be seen.

The Jurassic species were mostly small, some no larger than sparrows, and they flew by flapping their wings, probably more slowly and less powerfully than birds do. The breastbone was keeled for the attachment of flying muscles, but this feature was less strongly developed than in birds. They are generally supposed to have hawked insects in flight; dragonflies were abundant in the Mesozoic. Some may have snatched little fish from the surface of the water as modern terns do. They must have flown fairly well to be able to hunt in this way, but the small flapping pterosaurs were excelled by the birds in almost every available environment, and in Cretaceous times most of them had become specialized for a particular kind of flight, gliding and soaring. This led to the development of long, narrow wings and to a great increase in size, so that the big, late Cretaceous pterosaurs were the largest animals that have ever flown in the air.

One of them, the great *Pteranodon* of the Niobrara Chalk of Kansas has been the subject of more speculation and experiment than any other fossil animal. The aeronautics of its flight have been studied and models of it constructed and tested in wind-tunnels, and a reasonable suggestion has been made regarding the colour of its fur. Its total weight was 40 pounds (18 Kg) or less, its body the size of a turkey's and the span of its wings 23 feet (7 m). Its name means 'winged-toothless', and it had a long beak balanced by an elongated crest rising from the back of its skull. The location of some of its fossils shows that it ranged far out to sea, and it is pictured as leading the same sort of life as an albatross, but as being even more perfectly adapted for soaring and gliding, at the expense of any sort of progression on land, for which its spindly hind legs were quite unsuited. With a weight of 40 pounds, the wing-loading would have been about 1 pound per square foot, compared with 4 pounds per square foot in man-made gliders. This, combined with the enormous wing-span, enabled it to glide at speeds as low as 15 mph, so that a breeze of this strength would simply lift it off the ground like a piece of paper, or indeed off the water. It has been argued that it could never rest on water, but if it was covered with fur this could have been oil-impregnated, unwettable and buoyant just as a sea-bird's feathers are. It is likely that the fur was white, because slow gliding under a hot sun would lead to absorption of heat by such a great area of dark-coloured wing, and circulation of blood from the wings would overheat the body. The strange bony crest seems hard to account for in an animal whose every other characteristic is weight-saving. If it is admitted that the long beak was essential for catching fish, then the crest would act as a compensator for maintaining stability when the beak was lowered and also as a sort of weather-vane to stop the beak being blown to one side in the head wind created by forward movement. The neck muscles needed to provide for this would have consumed energy and weighed at least as much as the crest.

We can picture *Pteranodon* as spending most of its time at sea, gliding effortlessly and catching fishes at the surface. Its habitat ashore was probably offshore islands where it could rest and rear its young without molestation.

In all but the most recent discussions of *Pteranodon* palaeontologists are unanimous on one point: this must be the largest animal ever to have flown, nothing bigger could be airborne. But during the years since 1972 excavation in Texas has revealed remains of a pterosaur with arm bones twice the length of *Pteranodon's*. It has been named *Quetzalcoatlus*, and estimates of its wing-span range from a conservative 33 up to 50 feet (10 to 15 m). It lived far from the sea and its mode of life is believed to have been that of a gigantic pterosaurian vulture, soaring high in the sky watching for dying dinosaurs and descending to feed on their putrefying carcasses. If it required a wind to enable it to take off and become airborne it must have been in danger from predators when feeding on still days, for it was wholly defenceless. Or was it? If we knew skunks only as fossils we should have no hint or inkling of the fearful defensive weapon that they wield. The fragile soaring giant *Quetzalcoatlus* could have been armed in some comparable way that caused the fiercest predatory dinosaur to blench.

Reptiles of the Sea

The reptiles which dominated the surface waters of the sea in the Mesozoic fall into three main groups, the ichthyosaurs, the plesiosaurs and the mosasaurs. The first of these were the most highly adapted for marine life, analogues of the whales and dolphins of today. Their bodies were streamlined and fish-like. In some fossils the outline of the body has been preserved as a carbonaceous film, revealing a fleshy dorsal fin and a two-lobed vertical tail fin, the upper lobe fleshy, the lower supported by the backbone, which extends to its tip. All four limbs were present, developed as paddles, the fore pair twice as large as the hind. The pattern of bones seen in almost all land vertebrate limbs is much obscured. The upper arm and leg bones, humerus and femur, are very much shortened and all the bones beyond them reduced to a pavement of flattened discs, far more numerous than the normal bones of a vertebrate hand or foot. The jaws were long and usually toothed and the eyes big and surrounded by a ring of small bony plates, perhaps as a protection against changes in pressure during deep dives. The biggest of them were about 30 feet (9 m) long; 8 to 10 feet (2 to 3 m) was the more usual size.

The ichthyosaurs could not possibly have come ashore to lay eggs, and it is known that they were viviparous, that is they produced living young. Fossils of females containing well-preserved embryos have been found, and even specimens in which the embryo is partly emergent, fossilized in the act of being born. They fed on fishes and the squid-like cephalopods known as belemnites. Ichthyosaurs are found first in Triassic strata, were most numerous in the Jurassic and apparently died out before the end of the Cretaceous. This seems strange because the plesiosaurs and mosasaurs, apparently less well adapted for life in the sea, survived to the end of the period.

The ancestry of the ichthyosaurs is quite obscure, and it is possible that they evolved from the Amphibia independently of other reptiles, with viviparity taking the place of the amniote egg as an advance on the prodigal egg-laying of the amphibians. Whatever their pedigree, it is best to regard them as reptiles because they had reached a reptilian level of organization.

Plesiosaurs were descended from a group of Triassic reptiles called nothosaurs. These were elongated animals with feet adapted for swimming, but probably used for scrambling about on shore as well. In their evolution from these amphibious animals the plesiosaurs became strictly

OVERLEAF
Left The small Jurassic pterosaur *Pterodactylus* flying in at sunset to roost in a cave. Like the modern bats, pterosaurs probably hung by their feet when at rest.

Right These lifelike models show how pterosaurs must have appeared when flying across the sky. The small flock are *Pterodactylus*, the single one *Rhamphorynchus*.

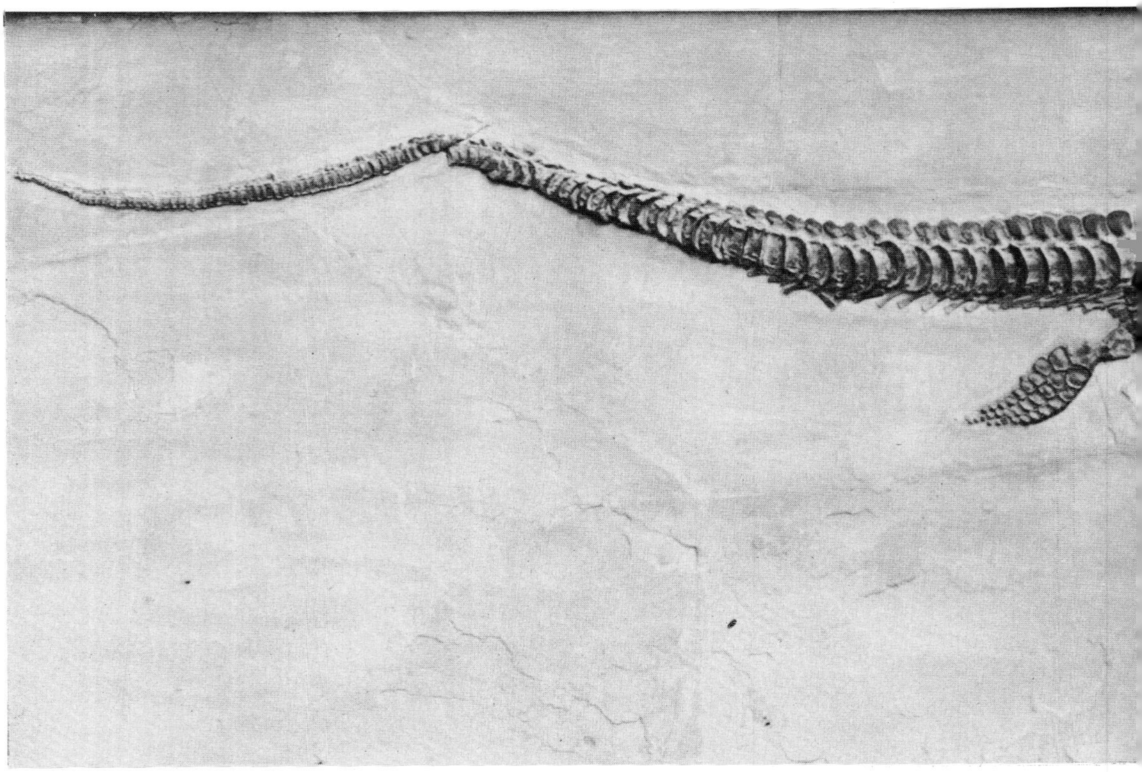

A fossil skeleton of *Ichthyosaurus* with that of an embryo preserved inside it.

Below The structure of ichthyosaurs. (a) The outline of the body as revealed by specimens in which it is preserved as a carbonized film. (b) The tail and tail-fin of a primitive ichthyosaur (*Mixosaurus*) from the Triassic. (c) The tail and tail-fin of *Ichthyosaurus* from the Jurassic.

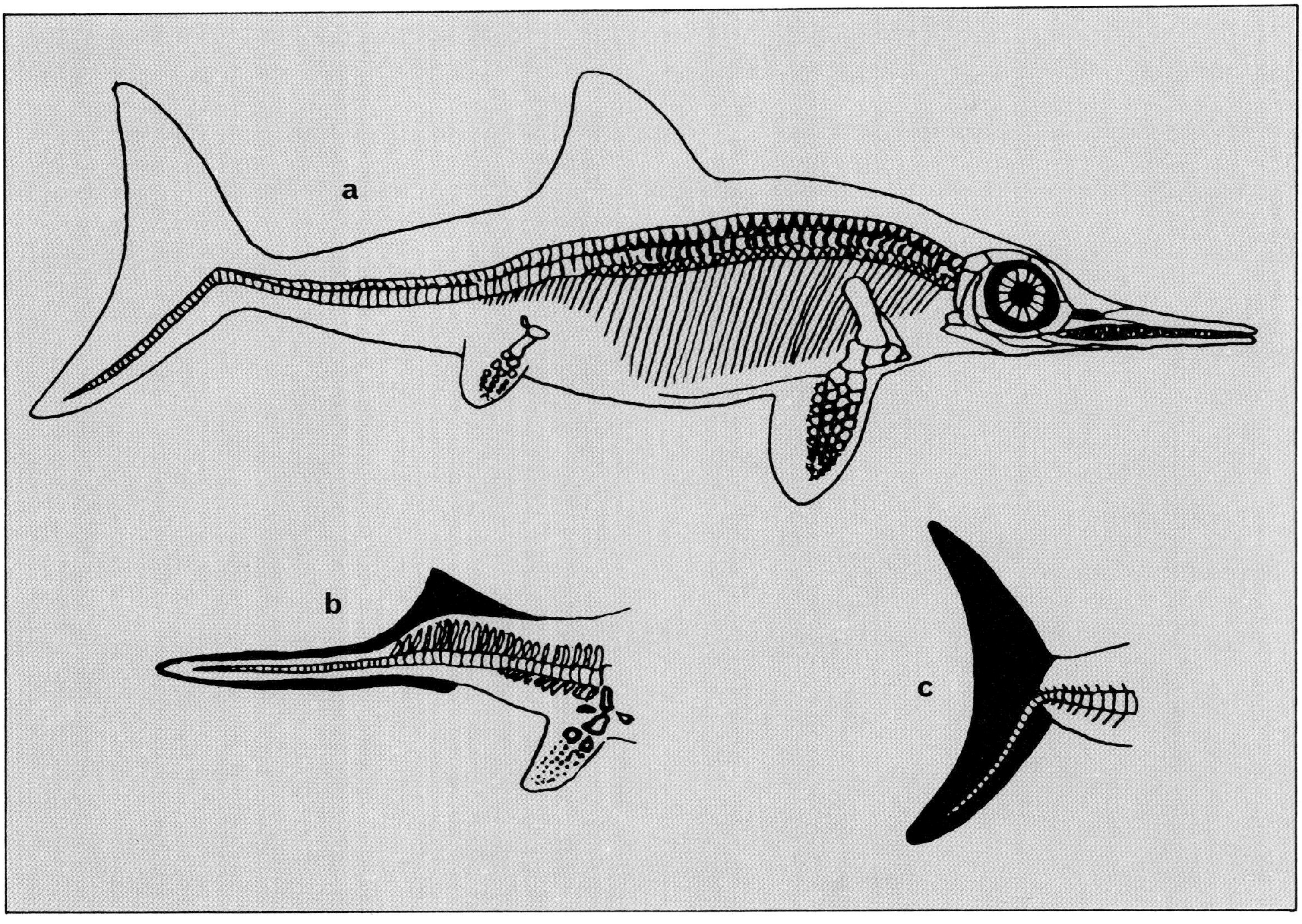

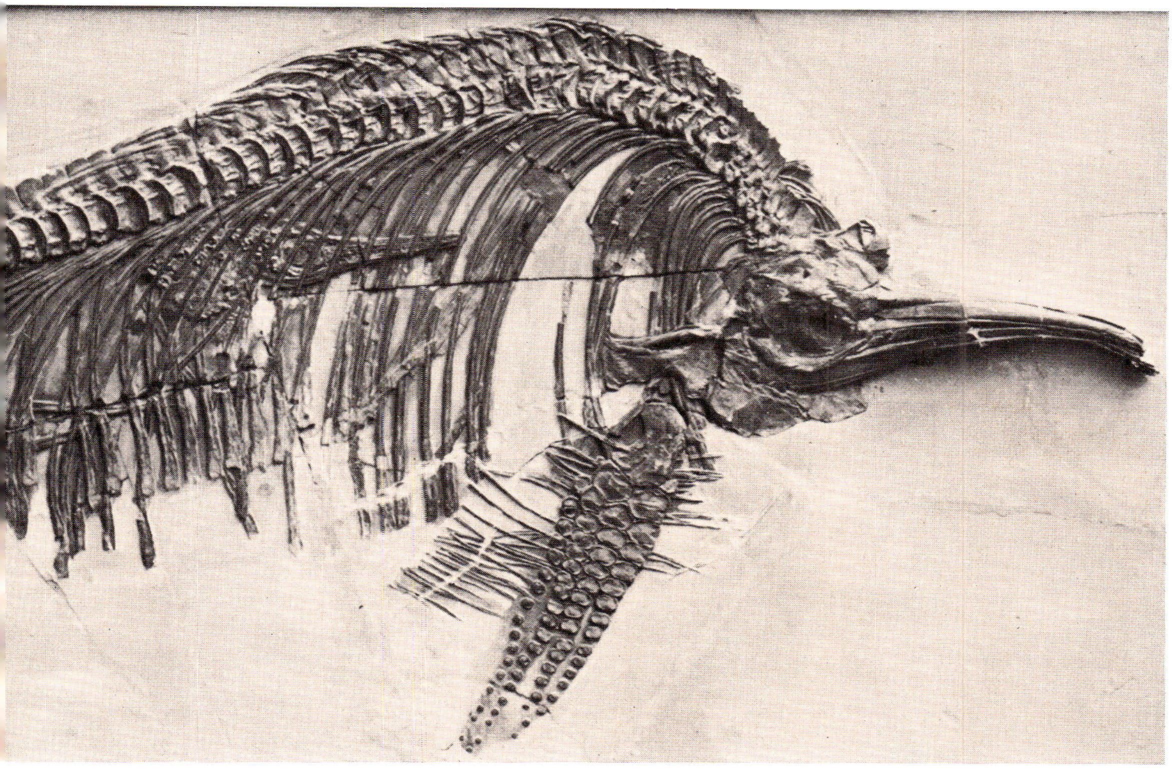

aquatic, swimming with their large paddles rather as turtles do. There is no direct evidence about their mode of reproduction, but they were probably viviparous.

Two distinct types of plesiosaur developed from these beginnings, the more familiar long-necked sort and the short-necked plesiosaurs or pliosaurs. The long-necked forms had a short rounded body, a long neck and small head, and they culminated in the Upper Cretaceous *Elasmosaurus*, which reached 40 feet (12.3 m) in length, at least half of which was neck. The pliosaurs had a very short neck and large head, which became relatively larger in the biggest kinds. The Australian Cretaceous pliosaur *Kronosaurus* was about 40 feet (12.3 m) in length, of which the skull accounted for 9 feet (2.7 m). A similar pliosaur called *Liopleurodon* comes from the British Jurassic. These were the largest of all the marine reptiles. Pliosaurs lived rather as sperm-whales do, diving to hunt large cephalopods, but the long-necked plesiosaurs lived at the surface, catching fish by twisting and turning and darting with their long necks. By interpretation of muscle scars on the bones we know that the pliosaurs made powerful backward strokes with their paddles, driving themselves forward, but the muscles controlling the forward or recovery stroke were fairly small. The fore paddles could also make the up-and-down movements associated with diving and surfacing. The long-necks, on the other hand, could make forward or backward horizontal strokes of equal power, enabling them to turn very quickly on the surface. If the Loch Ness monster were a surviving long-necked plesiosaur it would spend all its time at the surface and be seen by every tourist who drove past. Pliosaurs are now known to have had a small vertical tail fin.

The mosasaurs had a short history restricted to the later part of the Cretaceous. They were gigantic marine lizards closely allied to the existing monitor lizards. The body was elongate, almost serpentine in shape, with the tail flattened as an oar for swimming, and the limbs were short paddles with webbed fingers and toes. The jaws were very long and beset with strong conical teeth and the lower jaw was hinged in the middle making the gape very elastic, so that they could swallow large chunks of food. They were the most rapacious predators of the Upper Cretaceous seas and preyed on anything they could catch: a large ammonite shell has been found with distinct mosasaur tooth-marks on it. No fossils of young mosasaurs have ever been found, which is remarkable since remains of the adults are widespread in marine deposits. One suggestion is that the females ascended rivers to breed and the young spent their early lives in fresh waters, a more safe and sheltered environment than the open sea, but this is pure speculation.

Apart from the dinosaurs themselves we have looked at all the groups that played an important part in the history of the age of reptiles, and that became extinct at or near the end of

OVERLEAF

Above left A museum model of *Ornithosuchus*, a thecodont of the type immediately ancestral to the dinosaurs.

Above centre The fossil skull of *Heterodontosaurus*, showing the development of enlarged 'canine' teeth, a feature quite unique among dinosaurs.

Below left Excavating a fossil ichthyosaur skeleton. The series of vertebrae making up the tail of the animal has been exposed.

Above right The skeleton of an ichthyosaur in the course of excavation; plaster is being applied to minimize the risk of breaking the bones when they are removed.

Below right A stage in the excavation of the fossil skeleton of a plesiosaur.

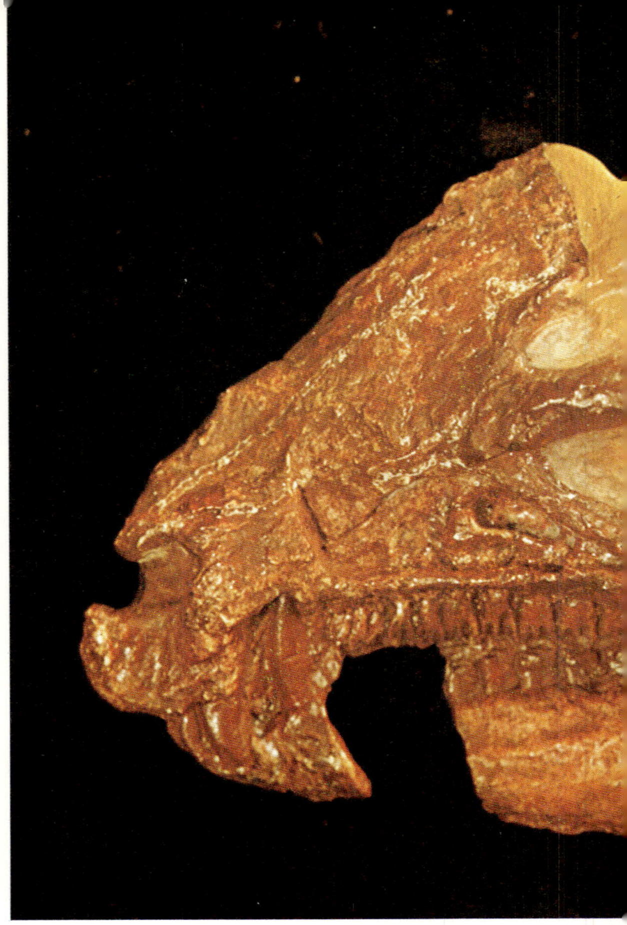

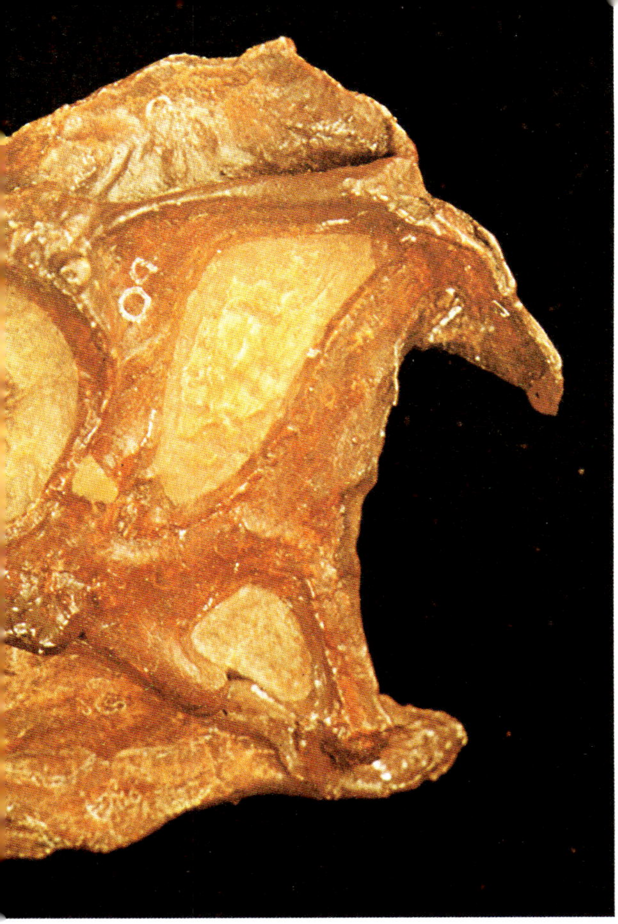

it. A few extinct Orders played a minor role, among them the rhynchosaurs, which flourished briefly in the middle Triassic. They were heavily built reptiles, as big as pigs or a little bigger, and strangely equipped for eating. They had a beak rather like a parrot's and on both jaws and palate numerous tiny teeth which constantly wore out and were renewed; they must have worked like a rasp and the food was presumably hard or fibrous vegetable matter of some kind. The remarkable tuatara of New Zealand is distantly related to them.

Among the marine reptiles the placodonts were also confined to the Triassic. They had an armour of bony plates on the back, a fairly long tail and webbed feet, and their teeth show that they fed by crushing up the shells of molluscs.

Except for the snakes, which had their beginnnings in the Cretaceous, all the modern groups of reptiles were well established by Upper Triassic times. Lizards evidently flourished through most of the Mesozoic but are rare as fossils, probably because they lived in the uplands. Fissure deposits in old limestone land surfaces are almost the only source of Mesozoic lizard fossils, as they also are of remains of the little dinosaur-shy mammals of the era. As early as the Upper Triassic a lizard called *Icarosaurus* had developed rib-supported wing-membranes and was gliding among the trees just like the elegant little *Draco* lizards of the Far-eastern rain-forests of today.

Crocodiles, the only surviving archosaurian reptiles, have existed without much change since they evolved from the thecodonts. In the Upper Cretaceous one of them rivalled, if it did not excel, *Tyrannosaurus* as the largest known land carnivore. This was *Phobosuchus*, the 'terror crocodile', which certainly ate dinosaurs for breakfast. Its skull is 6 feet 3 inches (1.9 m) long and its length is estimated at 45 feet (13.7 m) compared with about 30 feet (9 m) for the largest living crocodiles.

The Chelonia, the turtles and tortoises, first appear in the Triassic. Their skull is of a type similar to that of the cotylosaurs, the most ancient reptiles, and even in the time of the dinosaurs they were already relics of a bygone fauna. The biggest of them was *Archelon*, a huge marine turtle that swam in the seas of the late Cretaceous with the mosasaurs and the last of the plesiosaurs. It had a shell 10 feet (3 m) long, considerably larger than the 6 to 7 foot (2 m) leathery turtle of today. But the turtles have never changed very much. Their fossil history curiously parallels the well-known fable of the tortoise and the hare: in the last 200 million years the dinosaurs, ichthyosaurs, plesiosaurs and pterosaurs have run, swum and flown past and over the turtles, only to lapse into oblivion. It would be nice to think that they will still be plodding along when we have gone, but I fear that we shall destroy them in the course of destroying ourselves.

The Geography of Past Ages

The science of zoogeography concerns the distribution of animal types in the world as determined by their evolutionary history: their 'natural' range as opposed to anything arising from human agency, like the present distribution of horses and Colorado beetles. From studies in this science we have become quite accustomed to the idea of one sort of marsupial in Australia, another in South and Central America and no marsupials anywhere else at all; of sloths, anteaters and armadillos only in South and Central America; of giraffes only in Africa. This sort of distribution seems reasonable because the continents are to a great extent separated from each other by the oceans, and their faunas have evolved independently in isolation from each other.

Until quite recent years it was thought that the continents and oceans have always occupied their present positions. It is true that in 1922 Alfred Wegener assembled and published a great deal of evidence that, at a not very remote period of geological time, all the continents were assembled to form a single land mass which he called Pangea or 'all-earth'. He maintained that since that time Pangea had fragmented and the pieces had drifted, like gigantic icebergs, to their present positions. His theory was rejected by geologists because no mechanism could be imagined which would move the continental blocks through the rigid substratum in which they are embedded.

Evidence that Wegener was right continued to accumulate despite orthodox hostility to his theory, and within the last decade the science of plate tectonics has been elaborated and provides a satisfactory explanation of continental drift, which is now an accepted fact of geology. The details of plate tectonics are beyond the scope of this book, and so is the main movement of the continents to their present positions, because the greater part of this took place in Cenozoic times, after the end of the Mesozoic when the age of reptiles ended. We are concerned, however, with the disposition of the land masses from Permian to Mesozoic times.

During almost all of this 200-million-year period we have good evidence that the present continents were united into two supercontinents, a northern Laurasia formed of Eurasia and North America, now divided by the North Atlantic, and a southern Gondwanaland. The units that were once bunched together to form the southern supercontinent are now widely separated; they are South America, Antarctica, Africa, Australia and (rather surprisingly) Peninsular India. In Permian and Triassic times the two great land masses were probably broadly linked by a junction of North and South America with what is now the north-western bulge of Africa. To the east there was a deep embayment at the same latitude as the junction, but this junction was broad enough to justify speaking of a 'Pangea' at that time. In the Jurassic the eastern embayment encroached and a western one, the beginning of the split between North and South America, was developed. These two embayments led to the formation of the great Tethys ocean when they joined up and separated the supercontinents in Cretaceous times.

The zoogeography of the ancient reptile faunas could be expected to reflect the early arrangement of the continental masses. It does so in a way that has always lent strong support to the reality of continental drift. The most impressive item of evidence of this kind concerns the distribution of the small hippopotamus-like cynodont reptile *Lystrosaurus*, already mentioned earlier in this chapter. Fossils of this animal have been found in South Africa, in Bengal in India, in Singkiang in China and in Antarctica, indicating that it lived on all these areas of land (and certainly others where no fossils have been preserved or discovered) in early Triassic times. *Lystrosaurus* was not an animal with exceptional power of dispersal; it could neither swim across oceans nor fly nor even run very actively, since it was adapted for life in swamps. With the continents in their present positions its distribution is impossible to explain, but with two supercontinents connected with each other all the lands of the world were continuous and mutually accessible. *Lystrosaurus*, by migrating along rivers and over any swampy areas that formed during the several million years of early Triassic time, was able to be virtually cosmopolitan.

Climates of Past Ages

Reconstruction of world climates in times as remote as the age of reptiles is not an easy task. The first thing to be realized is that the climate in which man evolved and now exists is an abnormal one. We are living in an ice age, in which glacial and interglacial periods have been alternating during the last one to two million years. At present we are in an interglacial, but it is only ten thousand years since the glaciers and ice-sheets last retreated, and there is no reason to suppose that they will never return.

Ice ages seem to be a consequence of the formation of polar ice-caps. These are believed to come into existence either when the continents move so as to enclose a polar sea, like the present Arctic Ocean, or when large continental masses drift into polar areas and become glaciated because the snow on them reflects back the sun's heat and reduces the temperature, so that the snow accumulates faster than it melts. The Antarctic Continent now lies, of course, right over the south pole. When neither of these conditions exists the free circulation of oceanic water between the equator and the poles keeps the world climate stable and fairly warm.

The only other ice age that has occurred during the time of the fossil record was a prolonged and severe one at the very beginning of the age of reptiles, 300 to 250 million years ago at the transition from the Carboniferous to the Permian period. It was confined to the southern hemisphere and affected large areas of Gondwanaland, which then lay right over the south pole. The present distribution of glacial deposits from those times extends over areas in all the southern continents and north of the equator in India. If these continents are reassembled to form Gondwanaland, and the glacial deposits related critically to their periods of formation, the centre of glaciation seems to progress from one side of the supercontinent to the other during the 50 million years of this ice age. It was, of course, the drifting of the great land mass across the south pole that produced this effect.

During the earlier part of the long southern ice age Laurasia lay over the equator and the great tropical coal forests developed, probably the first extensive forests that the world had seen. Later, during the Permian, the climate of this area changed and vast hot deserts replaced the forests. The reptiles started their history in the forests and evolved to become the dominant land vertebrates on the habitable fringes of the deserts. *Dimetrodon*, well adapted to absorb heat after the cold desert night and radiate away excess heat in the daytime, was one of these. As Gondwanaland drifted away from the pole and its ice-sheets melted, it too became populated by primitive reptiles.

OVERLEAF
The ichthyosaurs were reptiles with fish-like bodies, perfectly adapted for life in the sea. They are shown here diving over a coral reef in pursuit of a shoal of fish.

By Triassic times there was no glaciation in the world, but arid conditions continued to be widespread. These were followed in the Jurassic and Cretaceous by a warm climate, wet enough to support abundant and varied vegetation world-wide over the two supercontinents. In this Utopia the dinosaurs reigned supreme and were the only large land animals.

The Evolution of Plant Life

I have made some references to the vegetation which was evoked by the climates that succeeded each other in different parts of the world. The plants, then as now, made the environment in which the animal life existed, and I will end this chapter with a brief survey of the floras that evolved together with the reptiles during their long mastery of the land.

The earliest reptiles lived in the Carboniferous coal forests, which flourished about 300 million years ago in the tropical swamps of the southern part of Laurasia; at this time Gondwanaland was in the grip of a great ice age. As is well known, the forests were dominated by giant relatives of the small modern plants known as club-mosses and scouring-rushes or horsetails. These are classified, together with the ferns, in the subdivision of plants called the Pteridophyta. Among the giant horsetails were the scale-trees, such as *Lepidodendron* and *Sigillaria*, with trunks 6 feet (1.8 m) in diameter and 100 or more feet (over 30 m) in height. The characteristic scale markings on their fossil bark are the scars of leaves which were attached to the stem and fell off as the tree grew. Ferns and tree-ferns were also abundant and not very different from those of today. Relatives of the ginkgo or maidenhair tree were well represented. The single surviving species of ginkgo is one of the most remarkable of all botanical 'living fossils'. The earliest seed-bearing plants also appeared in these forests. They are called seed ferns and had fern-like foliage but bore nut-like seeds instead of spores. These are the oldest of the gymnosperms, represented today mainly by the conifers. Some primitive conifers also appeared in the Carboniferous period.

During the following Permian period the ice retreated in Gondwanaland leaving a cool climate behind it. Here a totally different flora developed, not at all rich or diverse, and dominated by plants with large tongue-shaped leaves called *Glossopteris*, a member of the order of plants called seed ferns. In some places they were luxuriant enough to form coalfields. In the north the climate became hot and arid in the Permian and the coal forests dwindled away, but remnants of their plant types persisted in adequately watered havens. Primitive conifers and ginkgos replaced the scale-trees and giant club-mosses as the dominant plants. This type of flora became widespread in the Triassic as conditions became warmer in the south, but the dry climate persisted and Triassic plant fossils are rather rare in consequence.

The Jurassic flora is much better known. At this time ferns and other ancient plants were still common and so were trees of the ginkgo family, but the landscape was dominated by conifers resembling modern araucaria (monkey-puzzles), yews and cypresses. The cycads, a family of gymnosperms of which a few kinds still survive, were prevalent in the Jurassic. They look rather like short-trunked stumpy palm-trees, but of course are not related to palms in any way. The reproductive structures are cone-like, and pollen-bearing and seed-bearing cones are borne on separate plants. The cycadeoids or bennettitales had an appearance rather like cycads, but the male and female reproductive organs were combined to form a flower-like structure. They are not, however, believed to be ancestral to the true flowering plants. Cycadeoids were sometimes very abundant, making up a large percentage of the forest vegetation, and no doubt of the diet of the herbivorous dinosaurs. In spite of this abundance they became extinct before the end of the Cretaceous period. One of the signs of an equable climate is the absence of growth rings, which indicate contrasting seasons, in the fossil tree-trunks of Jurassic age.

The early Cretaceous woods were very like those of the Jurassic, with conifers, cycadeoids and ferns still the dominant plants, but flowering plants had appeared. Familiar types like magnolia, oak and laurel were present, and some tropical genera such as breadfruit and camphor tree. These probably evolved during the Jurassic in an upland environment from which no fossils have been preserved. By mid-Cretaceous times, 35 million years before the end of the period, the flowering plants were dominant and the landscape had much the same aspect as the natural forests of today. The last and some of the greatest dinosaurs lived in a setting that would seem in no way strange to anyone familiar with modern tropical savannah and rain-forest. The birds were already represented by familiar types and there were bees, and almost certainly butterflies and moths, pollinating the flowers, but fossils of insects are rare. Large mammals, of course, were lacking because the reign of the dinosaurs was not yet over; nor was there any sign at all that it was nearing its end.

Chapter Three

Natural History of the Dinosaurs

OVERLEAF
The nothosaurs were marine reptiles of the Triassic Period, allied to plesiosaurs and perhaps ancestral to them. They were probably able to leave the water and scramble about on the shore.

Opposite Structure of the pelvic girdle in dinosaurs. (a) The saurischian type found in the sauropods and theropods. (b) A typical ornithischian pelvis. (c) That of *Scelidosaurus*, illustrating the primitive ornithischian type lacking the forward prong. The small sketches provide examples of the dinosaurs concerned.

No doubt dinosaurs ranged over a great part of the Mesozoic lands, the hill forests and upland valleys as well as the swamps and river flood-plains of the lowlands. But the lowland forms are the only ones of which any sort of a clear record is available to us, because animals living in hilly country are seldom preserved as fossils. Even of the lowland species our knowledge must be limited to a small percentage of the total that existed. It must be realized that the quite modest number of dinosaur genera that have been described and named are not members of one fauna, but of a succession of faunas ranging over a period of time double that which has elapsed since the last known dinosaurs lived. To depict a *Stegosaurus* confronting a *Tyrannosaurus* is just as unrealistic as the comic-strip absurdity of hairy men being chased by dinosaurs into their caves; *Stegosaurus* was extinct many millions of years before *Tyrannosaurus* came into existence.

Nevertheless research, ranging over all the continents, has provided fossils from the various periods which do give us a fairly clear picture of the ecology of the lowland dinosaurs. It resembles that of the larger mammals in regions such as Africa, where these still exist in some variety: a great diversity of herbivores coexisting with predators that are far fewer, both as regards types and individuals. The larger mammals afford the best material for comparison because in terms of size they are comparable with the dinosaurs, the smallest of which were not quite as big as the smallest antelopes. Many of the larger ones were about the size of elephants or hippos, though of course the largest dinosaurs were far bigger and heavier than any existing land animals. There were no mouse- or rat-sized dinosaurs; this vertebrate biological niche was filled then, as now, by lizards and mammals.

'Dinosaur' is really a term of convenience without any standing in zoology. It is analogous to 'ungulate', a name formerly applied to all the hoofed mammals, before it was realized that the even-toed ungulates (pigs, cattle, goats, antelopes) and the odd-toed ones (horses, rhinoceroses, tapirs) had entirely distinct evolutionary origins and should be placed in separate Orders within the Class Mammalia. Similarly there are two distinct Orders of dinosaurs, which were distinguished after the general term 'dinosaur' was proposed for all of them at a time when very few kinds had been recognized. Recently it has been maintained that all the dinosaurs do have a common evolutionary origin subsequent to their separation from the thecodonts. However, there is hardly any proposition about them that is not disputed by some authority or other, and I shall maintain the conventional division into two separate groups, which forms the basis of their classification.

The most important difference between the two Orders is in the arrangement of the bones forming the pelvis. In the Saurischia this is similar to that of other reptiles, including the thecodonts; the Ornithischia have a pelvis resembling that of birds, and the two are often referred to as the reptile-hipped and bird-hipped dinosaurs. In both groups the hip socket is surrounded by three bones, the ilium above, the pubis below and in front, and the ischium below and behind. The main difference is in the pubis, which in the Saurischia projects as a forward and downard prong. In the majority of the Ornithischia it has two prongs, one projecting forwards and up, and the other backwards, lying below and close to the ischium, just as in birds. Until recently all known dinosaur pelves were of one of these two types, but ornithischian dinosaur fossils are rare in early Mesozoic strata, and none had been found with the pelvis well preserved of earlier date than late Jurassic. A find of a small armoured dinosaur, *Scelidosaurus*, in the early Jurassic of southern England, and some late Triassic fossils in South Africa, have revealed that the primitive ornithischian pubis lacks the forward prong, which must therefore be regarded as an accessory growth from a backward and sub-ischial true pubis.

The other peculiar feature of the ornithischian dinosaurs was a small bone at the tip of the lower jaw, which lacked teeth and was covered by a horny beak. In many of them the front part of both jaws was toothless and provided with a beak for plucking vegetation, which was then chewed by batteries of grinding teeth further back in the mouth. All the ornithischians were herbivores, and the modern herbivorous mammals, such as ruminants and rodents, feed in rather the same way. They retain the front teeth or incisors for plucking or biting off their food, then there is a toothless gap followed by grinders at the back of the jaws. For this way of feeding cheeks are necessary to retain the food in the mouth, and the mammals and bird-hipped dinosaurs are the only animals that ever developed them.

The saurischian skull had no specializations of this kind. Typically there were sharp teeth all round both jaws, or in the front of the jaws, and the mouth gaped like that of a crocodile. In some of the smaller saurischians, however, the teeth were lost and replaced by a beak, the result being an animal rather like an ostrich, with fore limbs and without any feathers. Both Orders include biped and quadruped forms, and all the carnivorous dinosaurs were biped saurischians.

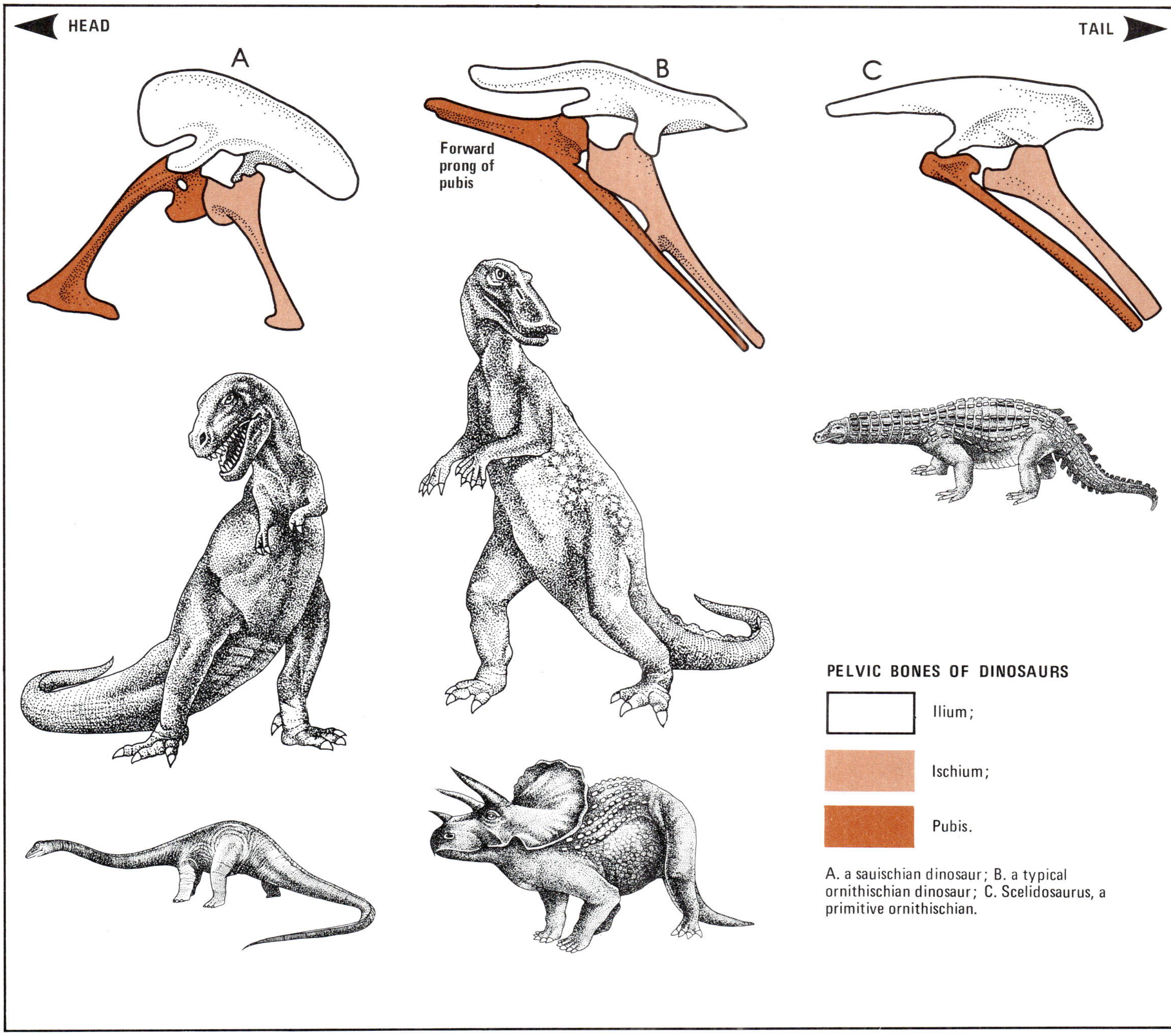

PELVIC BONES OF DINOSAURS

A. a sauischian dinosaur; B. a typical ornithischian dinosaur; C. Scelidosaurus, a primitive ornithischian.

The classification of the dinosaurs into suborders within the two Orders is shown on page 52. The descriptions and discussions that follow in this chapter will be based on this classification, beginning with the Saurischia, as they include the earliest known dinosaurs and show a transition from the ancestral thecodonts.

Opposite The rhynchosaurs, of which *Scaphonyx* is an example, were abundant for a time during the Triassic Period. They had a beak rather like a parrot's and numerous small teeth on jaws and palate. This was probably an adaptation for dealing with fibrous vegetable food such as roots.

Order Saurischia

Suborder Theropoda. The theropods are the carnivorous dinosaurs. A few of them may have been omnivorous, but they include all the dinosaurian flesh-eaters. They fall into two groups, the small, lightly built coelurosaurs and the large to gigantic carnosaurs. It is not a clear-cut division and there are some genera which cannot be certainly assigned to either; I will indicate these when we come to them.

We can now take up the story of dinosaur evolution where we left it on page 26, at the stage of the advanced biped thecodont *Ornithosuchus*. The next step is well illustrated by a late Triassic coelurosaur called *Coelophysis*, of which a remarkable find of perfect skeletons was made in New Mexico. This was a lightly built, slender animal with a long neck and very long tail, that ran on its hind legs. Its total length was about 8 feet (2.4 m), but it probably weighed no more than 40 or 50 pounds (20 kg); the whole of its structure seems to be aimed at reducing weight in order to attain maximum activity. The bones are all very delicate and the long bones are hollow like a bird's; the feet were also very bird-like. The hand had three functional fingers. The head was long and pointed and the teeth thin and blade-like with finely serrated edges. The pelvis was fully saurischian, with a long flattened ilium to which five of the sacral vertebrae were attached, providing the strong joint between the hips and the vertebral column needed for balancing and running on two legs. In the more primitive reptiles only two vertebrae were attached to the pelvis.

This was an animal that no large therapsid could catch and no small defenceless one escape, a perfect pattern of early dinosaur superiority over the mammal-like reptiles. It was a fierce little animal, not very fastidious in choosing its prey. In two of the most perfect skeletons that were found there were, within the body cavity, skeletons of much smaller individuals of the same species. At first sight this looks like evidence for a viviparous mode of reproduction, similar to that found for the ichthyosaurs. But they are rather large for unborn young and their bones are well-formed, not the cartilaginous bones usual in embryos. Most damning of all the evidence is the size of the pelvic opening of the adult *Coelophysis*, which is far too small to admit the birth of young of this size. Almost certainly it laid eggs, and rather small eggs at that. There seems to be no doubt that this little dinosaur was quite ready to hunt and devour its own partly grown young, and when it did so it swallowed them entire.

The smallest dinosaur yet discovered is *Compsognathus* from the late Jurassic of southern Germany, the size of a large fowl and very bird-like in structure. It had toothed jaws and probably hunted insects, small mammals and lizards. Its contemporary, *Ornitholestes*, was about 6 feet (1.8 m) long and very like a smaller version of *Coelophysis*. It was named 'bird-catcher' because it was imagined as preying on the very primitive birds that existed at the time. It was probably a general predator on small vertebrates.

In the Cretaceous period a group of remarkable coelurosaurs evolved that have come to be known as 'ostrich dinosaurs'. The best known of them are *Ornithomimus* ('bird-imitator') and *Struthiomimus* ('ostrich-mimic'), both from the late Cretaceous. They were as big as ostriches, or a little bigger, and remarkably similar to them in structure, even to the point of possessing a horny toothless beak. They differed most obviously in having a long muscular tail and well-developed fore limbs and, of course, in lacking feathers. A feature that they undoubtedly had in common with ostriches was the ability to run very fast. The arms were relatively longer than is usual in theropods, with long grasping fingers. These were probably used in hunting small prey and perhaps in pulling down fruit from trees. Very likely they were omnivorous, just as ostriches are.

A group of dinosaurs that may or may not be a branch of the coelurosaurs are the dromaeosaurs. They were small, very active bipeds, and by the end of the Cretaceous they had developed quite large brains that probably made them at least the equals in intelligence of the contemporary birds and mammals. *Stenonychosaurus* from the late Cretaceous of Canada is a recently discovered example of these very advanced 'mini-dinosaurs'. *Deinonychus* is an early Cretaceous member of this group. Its name means 'terrible claw' and it seems to have been a sort of dinosaurian panther, a formidable predator of moderate size, about 9 feet (2.7 m) long. It had two unusual features: the second toe of the hind foot did not reach the ground and ended in a

OVERLEAF
The long-necked plesiosaurs lived in the surface waters of the sea, swimming actively and capturing prey by darting movements of their necks. In the Cretaceous genus *Elasmosaurus*, shown in this picture, the neck was sometimes twice the length of the body.

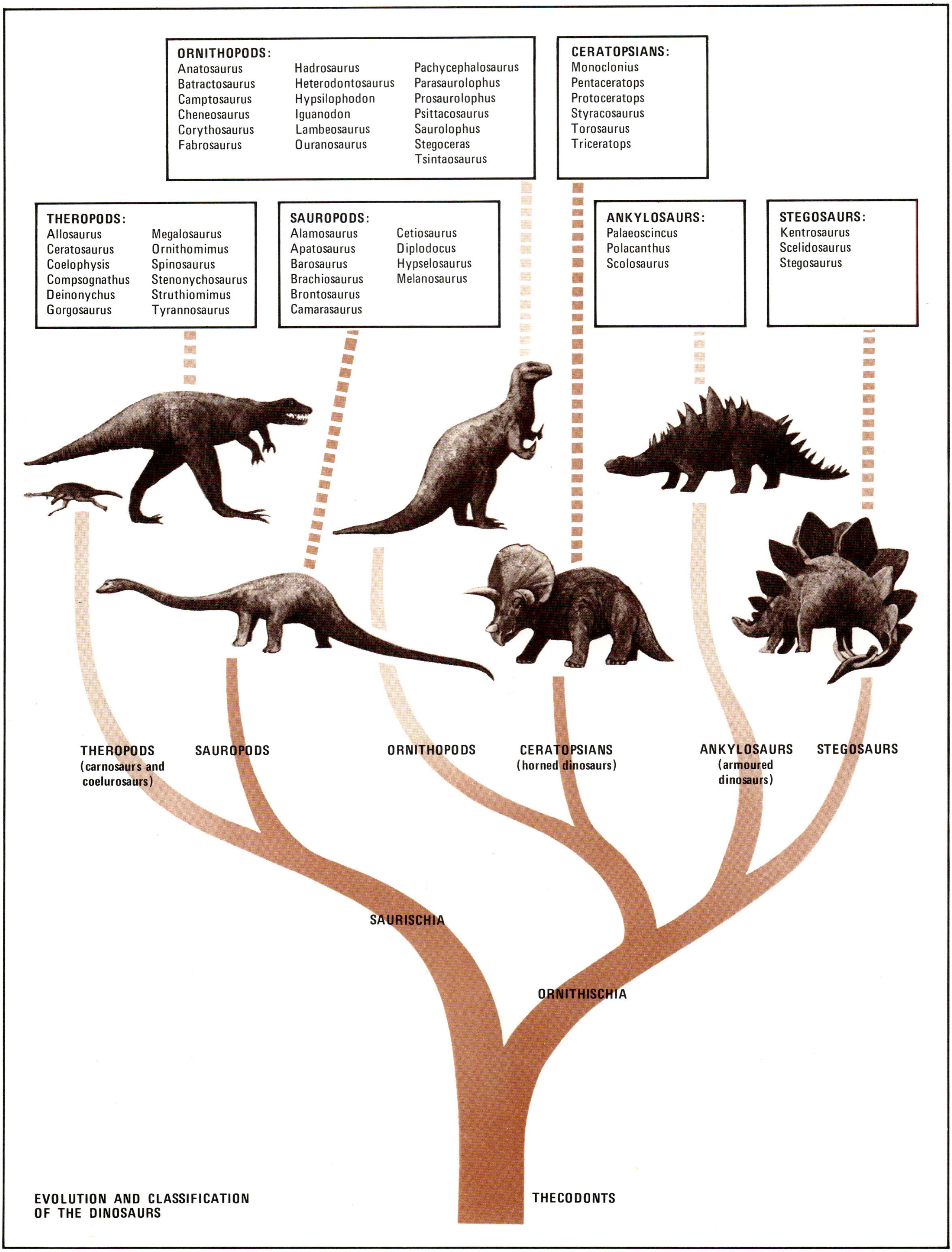

EVOLUTION AND CLASSIFICATION OF THE DINOSAURS

Opposite Diagram showing the evolution and classification of the dinosaurs. The names of all those mentioned in the text are referred to their respective Orders.

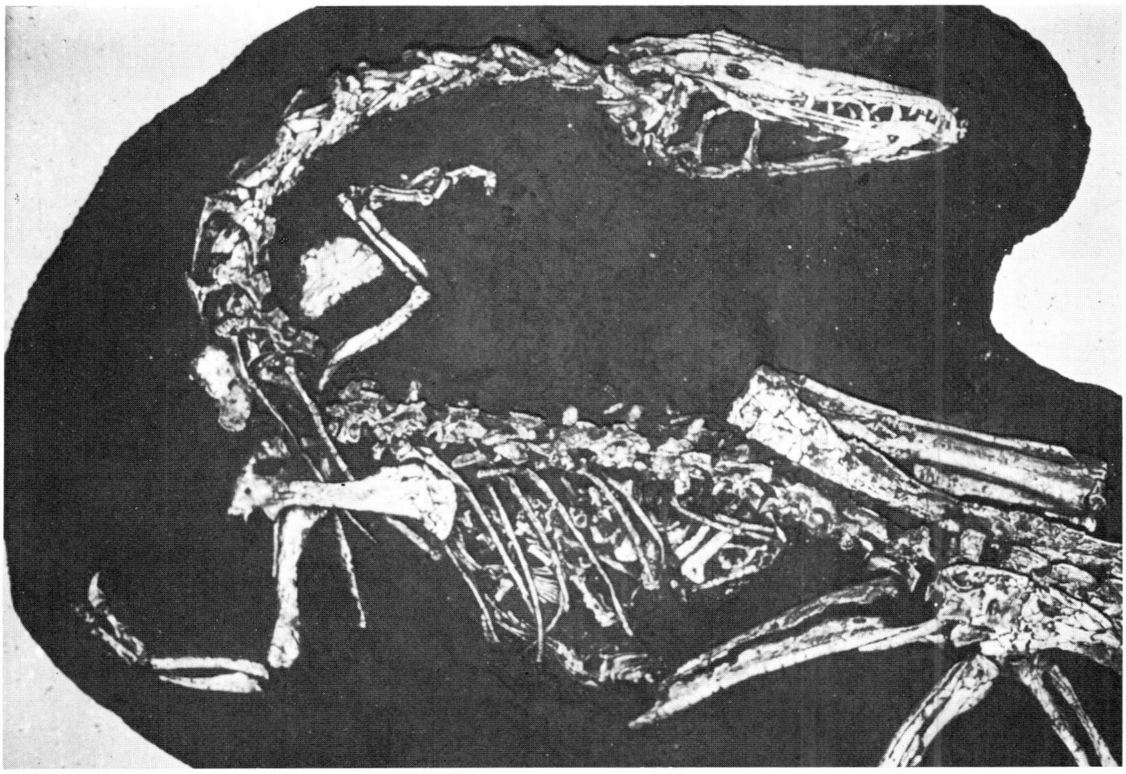

Left One of the fossil skeletons of *Coelophysis* found in New Mexico.

5-inch (13 cm) sickle-shaped claw, a fearful eviscerating weapon. Secondly, the long tail was sheathed in bony rods that must have held it stiffly horizontal, serving as a posterior balancing organ when the animal was running on its strong hind legs, or standing on one of them and slashing at the body of a victim with the other.

In one respect the coelurosaurs were the most successful of all the dinosaurs: they were the first to appear and they persisted right to the end of dinosaur time, constantly evolving new genera, never attaining any great size, always alert, agile bipeds. A recent well-supported theory maintains that not all the coelurosaurs did become extinct; one branch of them grew feathers and is still very much with us. I shall return to this subject in the next chapter.

The carnosaurs appear to be derived from coelurosaurian ancestors like *Coelophysis*, and to have diverged from them in one main characteristic, increase in size. They were the giant hunters of the Mesozoic, and culminated in the huge *Tyrannosaurus*, which in addition to being a palaeontological reality has become a sort of legend in contemporary folklore. *Megalosaurus*, from the English Jurassic and described in 1824, was the first dinosaur to receive a name, but its rather larger North American relative and contemporary, *Allosaurus*, is more completely known.

This was a formidable animal, over 30 feet (9 m) long, with a weight of 2 tons. It had immensely powerful hind limbs, three-toed, with large claws, on which it paced with six-foot strides. Its fore limbs, also bearing three claws, were much smaller and clearly useless for locomotion, but powerful enough to assist in subduing prey. The tail was thick, heavy and stiff, and balanced the dinosaur's weight so that it was poised easily on its hips, and it could run as fast as an active man, possibly faster. The head was large, with long jaws and numerous blade-like teeth. To lessen its weight the skull was reduced to a structure of bony arches with large openings between them, but it remained strong enough to support the action of the great jaw muscles. At the same time the bones at the back of the skull were loosely articulated to allow of some movement when large chunks of flesh, or entire small animals, were swallowed. Snakes have the skull modified in this way to an exaggerated degree. Although lightened as far as possible, the head was enormously heavy, and the neck accordingly short and thick, a contrast with the small heads and slender necks of the coelurosaurs. The brain, like that of almost all dinosaurs, was small.

Allosaurus was certainly a hunter and is actually known to have fed on the carcasses of the giant sauropods. The American Museum has a sauropod skeleton with the tail vertebrae scored by the teeth of an allosaur that was tearing the flesh off them. The marks exactly fit the teeth of a large predatory dinosaur of this type, and it even lost several teeth, which were found beside the

OVERLEAF
In late Cretaceous times, near the end of the age of reptiles, the fiercest predators of the sea were the mosasaurs, huge lizards wholly adapted for swimming. The giant marine turtle *Archelon* lived at the same time.

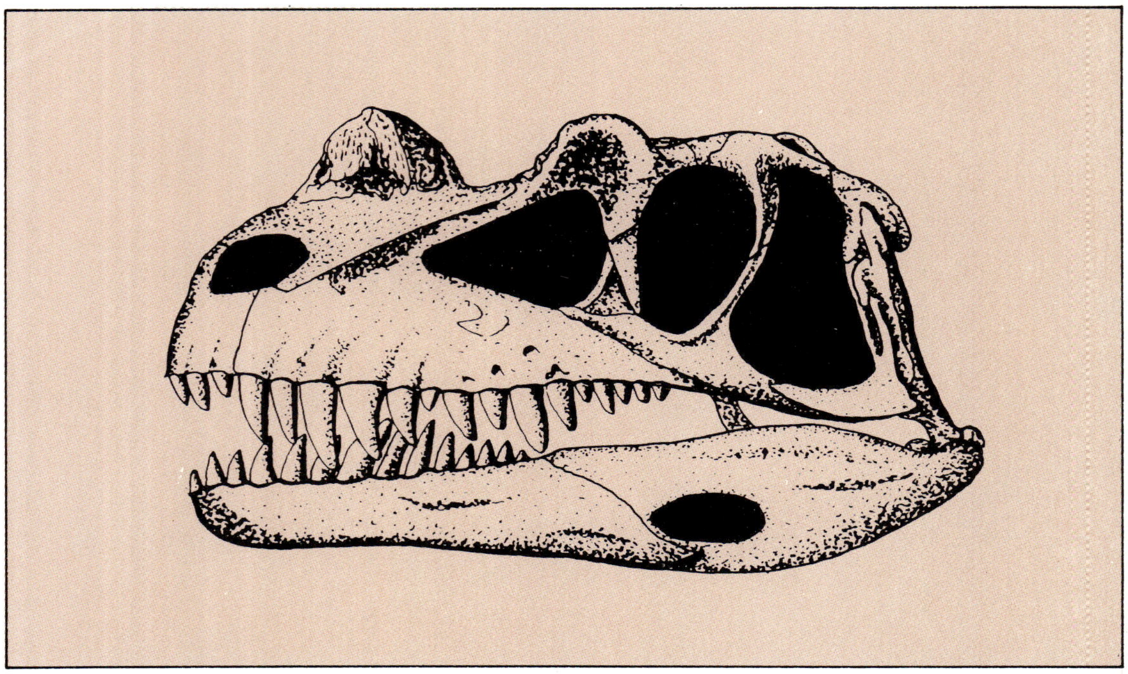

Detail of the skull *Ceratosaurus*. Its openwork structure is an adaptation to reduce weight.

sauropod's tail during excavation. One great advantage dinosaurs had over mammals was that they could replace worn or broken teeth indefinitely, and constantly did so, by the growth in the jaws of new teeth below the old ones. Whether the allosaur had killed its enormous victim, whether it did so alone or hunting in a pack, or whether on this occasion it was scavenging on a sauropod already dead, we can never say. We can be sure, however, that an allosaur could easily kill herbivorous dinosaurs up to two or three times its own size.

A smaller relative of *Allosaurus*, *Ceratosaurus*, had bony projections just above its eyes and a knob on its snout. *Spinosaurus* from the Cretaceous of North Africa, was larger than *Allosaurus* and had long spines sticking up from the vertebrae of its back, which must have supported a 'sail' recalling that of *Dimetrodon* (p. 20). This was probably a cooling device rather than a solar heater; dinosaurs were active animals, and large ones, with their relatively small surface area, would be more likely to suffer from internal overheating than from cold, unless this was much more severe than the vegetation of their period suggests. Heat would be dispersed when necessary by pumping blood through the arteries and veins in the fleshy tissue of the sail. *Spinosaurus* also had larger fore limbs than any of the other big carnosaurs. They were no doubt used as weapons but the animal may have walked on all fours when it was not in a hurry.

We now come to *Tyrannosaurus*, the only dinosaur that is commonly spoken of by its full binomial name, *Tyrannosaurus rex*. It is also commonly described as 'the largest carnivorous animal that has ever walked the earth', but there is no justification for saying this; it is the largest land carnivore known to us, that is all. The case of *Pteranodon* and *Quetzalcoatlus* should teach us to be wary of assigning superlatives in palaeontology.

Nevertheless *Tyrannosaurus* was an enormous and fearful animal. The tail was probably short (it has not yet been found complete), giving a total length of 35 to 40 feet (10–12 m). Reared up to its full height, it was 16 feet (5 m) tall and probably weighed over 8 tons, as much as two elephants. The skull was over 4 feet (1.2 m) long, shorter, by the way, than that of the terror crocodile, *Phobosuchus*. The teeth are 6-inch (15 cm) daggers with sharp serrated edges. The really strange and grotesque feature of *Tyrannosaurus* was its fore limbs, which were 30 inches (0.75 m) long and had only two fingers. It is difficult to see why the arms degenerated to this extent. One authority has suggested that they were used to help the animal to get up after it had lain down on its belly for a rest. Without the fore arms to brace it the body would just slide forward when the hind legs were straightened. But surely they would be more effective in this role if they were not quite so small. *Tyrannosaurus* must have had evolutionary ancestors with the arms of a normal carnosaur, why did natural selection lead to these mere vestiges of fore limbs? We do not know.

As a predator *Tyrannosaurus* must have been ponderous, with nothing like the speed of *Allosaurus*. For all its great size, footprints show that its stride was little over 3 feet (1 m). But there were ponderous herbivores for it to attack with its huge clawed hind feet and fearful armament of teeth. It lived right at the end of Cretaceous time, just before the dinosaurs' extinction. A near relative called *Gorgosaurus* lived a little earlier.

From stunted fore limbs we go to the other extreme in *Deinocheirus*. In 1965 an expedition to Mongolia found in Upper Cretaceous strata the shoulder-girdle and both fore limbs of a theropod with a three-fingered grasping hand. The fore arm alone was 8 feet (2.4 m) long! Nothing more is known about it, but even if the arms were disproportionately large (as indeed they must have been), *Deinocheirus* was a huge animal. When the rest of its skeleton is found it may well rival *Tyrannosaurus* in size.

Suborder Sauropoda. The sauropods have the saurischian type of pelvis and teeth either all round or in front of their jaws. Apart from these they had no obvious features in common with the theropods, because they had a totally different pattern of life. A clue to their ancestry is provided by *Plateosaurus*, which is sometimes placed, with some allied forms, in a small separate group, the Prosauropoda. *Plateosaurus* is a late Triassic dinosaur with a fairly long neck and tail, heavy body and the hind legs much larger and stronger than the fore legs. The degree of difference suggests that it normally walked on all fours but could rear up on the hind legs to reach up into trees and browse on the leaves. Some of the prosauropods may have been carnivorous, and they seem to provide a link between the wholly carnivorous and biped theropods and the sauropods, but not all authorities are agreed about this.

The true sauropods, often spoken of as 'brontosaurs', were the real dinosaurian giants, the familiar quadruped monsters with massive bodies, long necks and tails, and relatively tiny heads. They appeared right at the end of the Triassic period, flourished in the Jurassic and declined somewhat in the Cretaceous, though there were sauropods right up to the end of dinosaur time.

A number of genera have been described but on the whole they were uniform in structure, differing mainly in the length of neck and tail, relative to the body, and in the relative length and strength of the fore and hind limbs. Also, of course, they ranged in size from that of cattle to colossal animals of 80 tons and over. Usually the hind legs were longer than the front ones, but in the biggest one known, *Brachiosaurus*, the fore legs were longer and the neck very long, the two giving this enormous dinosaur an upward reach of 40 feet (12.3 m) and a length, in spite of the

OVERLEAF
Phobosuchus, the 'terror crocodile', lived towards the end of the age of reptiles and probably preyed on small and medium sized dinosaurs. It was much like a modern crocodile, but its skull is six feet long and it may have reached forty-five feet in total length.

The skeleton of *Ceratosaurus*, a large theropod with a horn on its nose.

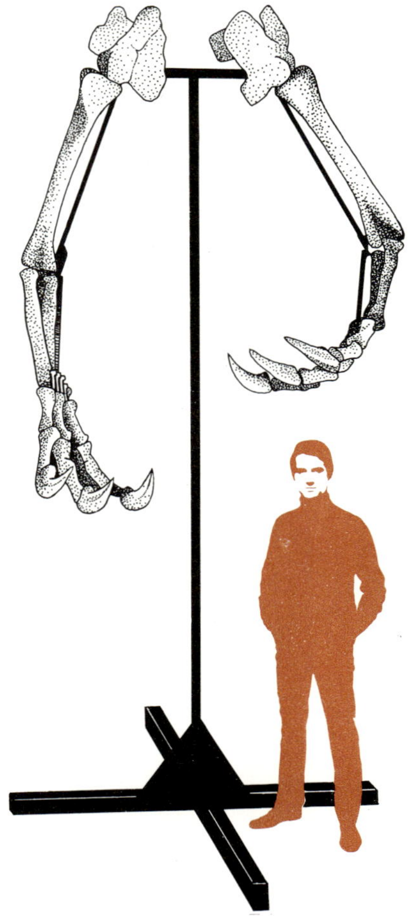

The fossil bones of the fore-limbs of *Deinocheirus*, with a human figure to provide a scale of size.

rather short tail, of 70 feet (21 m). The longest sauropod known is the much more lightly built *Diplodocus*, with very long neck and tail and a length approaching 90 feet (27 m), of which 45 feet (14 m) was tail and 26 feet (8 m) neck. This and *Apatosaurus* (formerly called *Brontosaurus*) are the two best-known sauropods.

The neck was always long or very long and was supported in life by a thick strong ligament running along the top of it; the neck vertebrae always have a large V-shaped notch on top in which the ligament lay. The head was disproportionately small, the jaws rather weak, with a uniform series of incisor-like teeth. In *Diplodocus* they are confined to the front of the jaws but in most sauropods they extend round a good part of them. The nostrils were often on top of the head; in *Brachiosaurus* they were raised up on a bony eminence on top of the skull. Some of them had the thumb of the fore foot armed with a large claw, otherwise the feet were huge padded extremities rather like an elephant's. Well-preserved footprints give us a good idea of what the feet were like.

I have mentioned a weight of 80 tons for the largest known sauropod, *Brachiosaurus*, and of over 8 tons for *Tyrannosaurus*. I am going now to make a short digression to explain how these weights are determined, for they are not mere guesses. The method was devised by the American palaeontologist Edwin H. Colbert. First a model of the dinosaur is made, on a reduced scale of course, as accurately as possible. The exact proportions of the articulated skeleton are reproduced and the requisite amount of muscle and other tissue is added, the muscle scars on the bones being used as a guide. The visceral cavity, as indicated by the ribs and the pectoral and pelvic girdles, is modelled, and the skill and flair of the experienced palaeontologist can be trusted to get the proportions right with only a small margin of error. The next thing is to find the exact volume of the model. As this is usually made of plaster it is not immersed in water (which would soak in and upset the result) but in sand. The method is to put it in a rectangular box and fill this with sand exactly to the top. The model is then carefully removed without spilling any sand at all, and the volume of sand needed to fill the box again must be that of the model. From the scale of the model to the animal itself the volume of the latter can be easily calculated; if the model was to a scale of length of 1:20 the volume must be multiplied by the cube of 20, that is 8000. All that is now needed is to estimate the specific gravity of a dinosaur. Dr Colbert used an alligator to determine this and got a figure of 0.9. As the figure does not vary

The skeleton of *Diplodocus*, a sauropod with a total length approaching 90 feet (27 m). Most of this was neck and tail and it was one of the less heavily built sauropods.

significantly from this in living vertebrate animals, it is entirely reasonable to apply it to dinosaurs. With volume and specific gravity known, the weight can easily be calculated, and Dr Colbert came up with some surprising results. The figure for *Diplodocus* was unexpectedly small, only 10 tons, but, as already mentioned, the specimen of *Brachiosaurus* on which the requisite model was based weighed 80 tons when alive. However, this skeleton does not represent the maximum size of *Brachiosaurus*; isolated bones have been found considerably larger than the corresponding bones of the complete skeleton. It may well be that the largest individuals of this dinosaur weighed as much as 100 tons.

Nothing remotely resembling a sauropod exists today so we have only their fossil remains, skeletons and footprints, to guide us in trying to determine how they lived. Opinion about this has diverged in two directions.

An animal that grows to a very large size encounters serious mechanical problems if it lives on land. Consider a fairy-tale giant 18 feet (5.5 m) tall, three times the height of a man. The area of the cross-section, that is to say the thickness and strength, of his muscles and bones, would increase as the square of his increase in height, so that he would be nine times as strong as a man. But his weight would increase as the cube, so that he would weigh twenty-seven times as much. If he were proportioned like a man he would be unable to walk or even stand up. He would, however, suffer very much less disadvantage in swimming because his great weight would be evenly supported by the water. By the same token a land vehicle the size of a battleship or oil-tanker would be quite unworkable. So, according to the earlier students of sauropods, would a land animal that approached a large whale in size.

When they first became known the sauropods were believed to be wholly aquatic and the earliest to receive a name was called *Cetiosaurus* or 'whale-reptile'. Subsequent discoveries made it clear that they were adapted for walking at least as well as for swimming, but their enormous size still persuaded zoologists that they must have spent almost the whole of their lives in the water, perhaps coming on shore at long intervals to lay their eggs, just as the wholly aquatic modern marine turtles do. A picture became established of sauropods swimming and wading in swamps, estuaries or coastal waters. Their food was supposed to be seaweed or freshwater aquatic plants, and in fairly deep water they would stretch up their long necks to breathe at the surface. The position of the nostrils on top of the head would seem to support this idea.

OVERLEAF

Left The short-necked plesiosaurs or pliosaurs included the largest of all the marine reptiles. *Kronosaurus*, found as a fossil in the Cretaceous of Australia, had a skull nine feet long and a total length of forty feet. It probably fed on large squids, diving deeply to hunt and capture them.

Right *Coelophysis* was one of the early carnivorous dinosaurs, a fairly small slender animal that ran actively on its hind legs. It provides a pattern of the sort of animal that led to the supremacy of the dinosaurs for many millions of years.

In recent years this picture of the sauropods has been effectively challenged. In the first place the idea of snorkel breathing with the body submerged to ten feet or more has been shown to be mere fantasy. The water pressure at even this moderate depth would make it impossible for the animal to inflate its lungs. Secondly the notion that large sauropods could not walk about on the land has been shown to be ill-founded, both by the evidence of deeply impressed footprints and by consideration of their skeletal anatomy. The school of thought which maintains that sauropods were primarily terrestrial animals has as its protagonist the American palaeontologist Robert Bakker. He argues as follows:

1. The limbs of sauropods are pillar-like and the rib-case deep in form like that of an elephant, and quite unlike the barrel-shaped body of the amphibious hippopotamus. The vertebrae of the back are strengthened by extra ligaments and articulations against gravitational strains.
2. The tail never has any sign of the side-to-side flattening which it would have developed in a habitually swimming animal, to serve as an oar or rudder.
3. The feet are not expanded for counteracting sinking in marshes; on the contrary the toes are short and stumpy like an elephant's. The footprints, too, indicate elephant-like feet.
4. The teeth are rather like the front teeth or incisors of browsing mammals and are sometimes severely worn as if used for tearing off dry fibrous vegetable food.
5. Sauropod fossils are found in river flood-plain deposits together with those of undoubtedly terrestrial dinosaurs.
6. There are existing land reptiles with dorsally placed nostrils, and also nostril openings high on the skull may indicate the presence of a proboscis, as in elephants and tapirs. The internal nasal organ in sauropods appears to have been large, suggesting a well-developed sense of smell, which aquatic air-breathing animals do not need.

Bakker pictures the sauropods as trampling through the forests of the river plains, reaching up to browse on the foliage of trees at heights far above the reach of other large herbivores. This would give access to a copious supply of food extending vertically to heights of 40 feet (12.3 m) for the tallest sauropods as well as far and wide over the landscape. The leaves would be stripped off by the small teeth (constantly replaced as they wore down and dropped out) and swallowed into a muscular gizzard well charged with stones which would mash the food up by churning movements of its walls. Birds swallow grit for just this purpose, and polished gastroliths or stomach stones have been found in association with at least one sauropod skeleton, and with those of other herbivorous dinosaurs. There is evidence that sauropods went around in herds, and these would have had a profound ecological effect in opening up the Mesozoic forests and allowing ground-level vegetation to grow, which would support the numerous other dinosaurs whose heads were close to the ground.

On the other hand, sauropods definitely did swim and wade in fairly deep water. A famous trackway of their footprints shows a succession of shallow impressions of the fore feet alone, as if the animal was swimming just within its depth, front feet lightly impressing the mud, hind legs and tail trailing. At the point where a single hind foot impression appears the track changes direction, showing that the sauropod turned by kicking the bottom with one hind leg.

There is no reason why the sauropods should not have been amphibious, just as the hippopotamus is. Hippos spend the day in the water and come out and feed on land at night. They are big animals, but not big enough to need a skeleton specially designed to cope with gravitational stresses during their landward excursions. They swim well, but there is very little in a hippo skeleton that indicates adaptation for swimming.

Let us imagine the sauropods spending the day (or night) feeding like giraffes in the forest and the other period swimming and wading in the wide rivers and estuaries of their habitat. They would wade and rest with the back awash and the neck outstretched on the surface, breathing easily through their dorsally placed nostrils. They would be able to rest comfortably with their great weight supported by water, and there is another reason why they might have lived in this fashion. Their great size makes it certain that they maintained their insides at a more or less constant temperature, high enough for active living; that is, they were warm-blooded (more about this in the next chapter). These huge animals, trampling about among the trees and reaching up with their thick muscular necks to feed, must have expended an enormous amount of energy which would generate internal heat. Owing to their very small surface area relative to their mass (see next chapter again), this heat would escape only slowly and would accumulate, leading to a risk of internal overheating. Water, with its high specific heat, is the most effective of all readily available cooling agents, as designers of nuclear power stations well know. So, by imagining the sauropods walking about, feeding and heating up for half of the twenty-four

Restorations of dinosaurs drawn in the first half of his century. *Above Cetiosaurus* by Neave Parker, *below Laelaps* (*Dryptosaurus*) by Charles R. Knight.

OVERLEAF
The 'ostrich dinosaur' *Struthiomimus* lived in late Cretaceous times and was remarkably like an ostrich in bodily structure even to the point of possessing a toothless beak. Its build suggests that it could run very fast, just as ostriches do today.

hours, and resting in the water, digesting their meal, drinking and cooling down for the other half, we arrive at a reasonable compromise between the old and the new interpretations of their way of life. The body temperature of the camel slowly rises during the heat of the desert day and slowly falls during the cold night, a fluctuation similar to that suggested for the sauropods, though quite differently caused.

The sauropods' most likely mode of defence against the giant predators was rearing and trampling with the fore feet, which were armed with a claw in some genera. If a theropod dared to attack in five or ten feet of water it would not only be trampled down but drowned as well. The notion that sauropods lashed their tails in self-defence is open to the objection that this would be ineffective both in a forest of tree-trunks and in the water.

As a sort of postscript to the sauropods I must mention a recent discovery made in Colorado of the pelvis, scapulae or shoulder-blades and five vertebrae of a hitherto unknown member of the group that must have excelled even *Brachiosaurus* in size. The largest of the vertebrae, which are from the neck, is 5 feet (1.5 m) long. Very little information about this find has been published, and the actual dethronement of *Brachiosaurus* can wait until the Colorado giant is formally described and named.

Order Ornithischia

The ornithischian dinosaurs include no carnivores and no really gigantic forms; the biggest of them weighed about twice as much as an elephant. They are perhaps less spectacular than the big saurischians, but they are far more diverse and some of them must have been extraordinary-looking animals. They all had in common the characters of a bird-like pelvis and a predentary bone at the 'chin' or tip of the lower jaw, carrying no teeth and covered by a horny beak. The front of the upper jaw was also usually toothless and beaked; only some of the more primitive ornithischians have teeth on the premaxillary bones. As in the saurischians, both quadruped and biped forms existed.

The descent of the saurischian dinosaurs from the Triassic thecodonts is well shown by fossils that can be arranged in a graded series between the two. The ornithischians were undoubtedly descended from thecodonts, but we do not know what their immediate thecodont ancestors were like.

Suborder Ornithopoda. This suborder comprises the biped ornithischians, and during the Jurassic and Cretaceous they were the most abundant dinosaurs. The name means 'bird-footed', but is rather misleading, as the coelurosaurs, the small, lightly built carnivores, had feet that were much more like those of birds. The fore limbs of the ornithopods were small, but never greatly reduced (as they sometimes were in the theropods), and these dinosaurs probably moved about on all fours when seeking food near the ground and only got up on their hind legs to stride or run or to reach up into trees. The tail was big and heavy and no doubt served as a counterbalance when running, as it did in the theropods.

The earliest ornithopods known come from the late Triassic of South Africa, and of these *Fabrosaurus* is the best-known. This was a small dinosaur, 3 feet (1 m) or so long, with long hind legs well adapted for running. Its upper jaw had the primitive feature of teeth in the front as well as at the sides; the front of the lower jaw was, of course, toothless. Another ornithopod, contemporary with *Fabrosaurus*, was *Heterodontosaurus*, also quite a small animal. It possessed the unique feature, for a dinosaur, of having in both upper and lower jaws a pair of long sharp teeth developed just like the canines of mammals. Its cheek-teeth show it to have been a herbivore, and the 'canines' may have been used in defence.

Another primitive ornithopod, small and with teeth in the upper jaw, was *Hypsilophodon*, which lived much later, in the late Jurassic and early Cretaceous – an early type of dinosaur surviving as an example to the numerous large ornithopods of the time of what their remote ancestors were like. A fancied resemblance to the modern tree-kangaroos persuaded some palaeontologists that *Hypsilophodon* was aboreal, and it is often portrayed perched rather awkwardly on the branch of a tree. The bones of its foot and lower leg were very long, and this is the mark of a runner rather than a climber; it was probably a fast, agile ground-dweller.

We now come to a typical large ornithopod, one of the best-known of all dinosaurs and the first one to be at all adequately described. This is of course *Iguanodon*, whose teeth puzzled Dr Gideon Mantell and his colleagues in the 1820s, and of which a specular find of about thirty skeletons, more than half of them entire, was made in 1878 at Bernissart in Belgium.

Iguanodon was an elephant-sized dinosaur, 4 to 5 tons in weight and able to rear up to a height of 14 feet (4.3 m). It could crouch on all fours to eat low-growing vegetation, stand up to browse on trees and run with head and body leaning well forward, balanced by the outstretched tail. Bones of *Iguanodon* are not uncommon in the Lower Cretaceous strata of south-eastern England, and the large three-toed footprints that sometimes come to light on quarry floors can safely be attributed to it. The fore limbs were much smaller than the hind and the fingers ended in blunt hoof-like nails. On the thumb there was a bony spike protruding at right angles to the fingers, and probably covered by a horny sheath in life. It is usually regarded as a defensive weapon but no one has explained how it could have been effective against the great claws and rending teeth of a large theropod. There are many things concerning dinosaurs that the fossil record can never tell us about. If these living animals were known to us only as fossils, who would be bold enough to suggest that the spur on the hind leg of the platypus or the spine on a sting-ray's tail were weapons charged with venom? Dinosaurs must have been very diversely adapted animals, and it is reasonable to suppose that most of the more obvious defensive devices seen among modern animals were evolved by them.

The great 'treasure' of complete iguanodon skeletons found at Bernissart was discovered in a coal-mine – not, of course, in the Carboniferous strata from which the coal was obtained, but from a deep infilled cleft in them that was an open ravine in early Cretaceous times, into which the dinosaurs evidently fell to their deaths. They are all mature, but include smaller and larger individuals, probably males and females. All have the spike on the thumb, so this is not likely to have been a sexual character to aid the male in holding the female in mating, which is one of the suggestions put forward to explain it. There are two theories of how they got there. The fact that the skeletons are vertically distributed through 150 feet (45 m) of flood deposits washed into the ravine suggests that this was a natural dinosaur-trap, operating over a long period during which it slowly filled up with silt. But in that case why did it catch only iguanodons and no other dinosaurs? The more dramatic notion is that a herd of them, possibly pursued by a carnosaur, stampeded and fell down the ravine's precipitous side, lodging at various levels on its lower slopes. If this is correct the absence of any young individuals is interesting as it suggests that they did not herd with the adults.

A possible ancestor of *Iguanodon* is seen in the Jurassic genus *Camptosaurus*. It is smaller and lacks the thumb-spike, and has been found in quantity in North America, though there are some European specimens.

Ouranosaurus lived rather later than *Iguanodon* and its fossil remains have been found in the same late Cretaceous formation as those of the theropod *Spinosaurus*. Like the theropod it has spines projecting up from the vertebrae, which must have supported a similar fleshy 'sail'. This seems to be an interesting case of two unrelated types of dinosaur adapted in the same way to cope with some hazard in the environment which both inhabited. Most likely this was the danger of overheating of the body in a climate unusually hot even in that period of widespread warmth. In *Ouranosaurus*, just as in *Spinosaurus*, blood could be allowed to circulate through the vascular tissue of the sail, which would act in the same way as the radiator of a car. At night and in cool spells of weather the flow of blood in the sail would be restricted. Any very large land animals living in warm climates will encounter this hazard. Elephants are believed to use their big flapping ears as heat-radiators.

In Cretaceous times a group of ornithopods called the hadrosaurs or duck-billed dinosaurs evolved, probably from iguanodonts of some sort. They were very diverse and numerous genera have been named, some of which include what must have been very strange-looking animals.

The oldest known hadrosaur, *Batractosaurus*, comes from the Lower Cretaceous of Mongolia, but they are abundant only in the Upper Cretaceous, especially that of western North America. Strata of the same age in central and eastern Asia have also yielded them in some variety. With the possible exception of some late Cretaceous fossils in South America they are unknown in the southern hemisphere. They were mostly large dinosaurs and none were light and small like *Hypsilophodon* and the coelurosaurs.

The best-known and one of the largest is *Anatosaurus*, about the same size as *Iguanodon*. It lived right up to the end of the Cretaceous, but was an unspecialized 'conservative' type of hadrosaur and will serve well as an introduction to the group. Its posture and gait were much as in the other large ornithopods, quadruped except when running or reaching upwards. The foot was three-toed, with hoof-like nails, the arm fairly long and strong with no little finger and a much reduced thumb; two of the remaining three fingers had small hoofs. The tail was flattened from side to side. The front of the skull and lower jaw are elongated and broadened to support the flat, horny, duck-like beak that gives the group its name. There were no front teeth above or below and the cheek-teeth were developed in a remarkable way.

OVERLEAF
Deinonychus and *Hypsilophodon* belonged to different orders of dinosaurs, but both were fairly small active bipeds. *Deinonychus* was a carnivore and had a claw on its hind foot specially adapted for slashing and disabling its prey. *Hypsilophodon* was a defenceless herbivore whose only hope of escape was to run and dodge even faster than its pursuer.

The skeleton of the hadrosaur *Anatosaurus*. The bones of the tail show clearly that it was flattened for swimming.

Each upper and lower jaw contained a battery of grinding teeth arranged in alternating rows and tightly packed together. The teeth are prism-shaped, and the outer side of each upper tooth and the inner side of each lower one are coated with enamel. The exposed teeth formed pavements in each jaw, and the differential hardness of the enamel and dentine caused them to wear unevenly, so that the pavements formed rough grinding surfaces. Under the exposed and functioning teeth there were others constantly growing up to take the place of those that wore down to the root and dropped out. Each jaw contained up to six hundred growing and functioning teeth, so that *Anatosaurus* and other hadrosaurs could have over two thousand teeth present in their heads at one time; no other land animals are known to approach this figure. The whole apparatus is clearly designed to grind up vegetable food.

The eyes were large and the presence of well-developed optic nerves and optic lobes in the brain indicates that *Anatosaurus* and its relatives had a keen sense of sight. There is a notch between the squamosal and quadrate bones of the skull, indicating a large eardrum. In another hadrosaur, *Corythosaurus*, the inner ear-bone or stapes has been found to be well developed. These dinosaurs had a good sense of hearing too.

Our knowledge of *Anatosaurus* is enhanced by the existence of two remarkable fossilized 'mummies' found in Wyoming many years ago. One is now in the American Museum of Natural History in New York, the other in the Senckenberg Museum in Frankfurt. The animals must have died in an arid region and the carcasses dried and shrunk instead of rotting and falling to pieces. The dehydrated mummies were then quickly covered by sediment, probably carried by a sudden flood, and buried before the skin softened. The fine silt and clay moulded the skin so that all its features are preserved, impressed in the rock. The skin is seen to have been closely covered with small lumps or tubercles and interspersed clusters of large flat tubercles, especially on the dorsal surface. It is claimed that webs of skin are present between the fingers and a crest of skin down the mid-line of the back and tail, but these features are not very clear and are disputed by some observers.

Anatosaurus belongs to the group known as flat-headed hadrosaurs in which, apart from the beak, the skull was not modified in any remarkable way. The others, the crested hadrosaurs, had extraordinary outgrowths, very diverse in the different genera, from the anterior bones of the skull, the premaxillary and nasal bones. These are again divided into the solid-crested and the hollow-crested hadrosaurs.

The first stage of growing a crest is seen in *Prosaurolophus*, in which the nasal bones extend back to meet the frontal bones, forming a low crest above the eyes. In *Saurolophus* the nasal bones extend back as a long upwardly curved spike over the top and back of the skull. *Tsintaosaurus*, a Chinese hadrosaur, had a crest formed in much the same way, but it was straight and projected vertically up, or even inclined forward, from above the eyes. The surface of the bone forming the crest does not indicate a covering of horn, but rather of skin, so it is unlikely that it was a weapon.

The hollow-crested hadrosaurs also show a gradation, from *Cheneosaurus*, with a low hollow crest on the skull, formed from the nasal and premaxillary bones, to a variety of fantastic forms. Chief among them are *Corythosaurus*, which had a helmet-like crest, *Lambeosaurus*, in which it was hatchet-shaped with a sort of accessory spine pointing backwards, and *Parasaurolophus*, in which the crest was a long curving tube, one and a half times the length of the skull. In the biggest species of this genus the length from snout to crest-tip was over 6 feet (2 m). In all these the nasal passage, running from the nostrils to the throat, loops up into the crest. The loop is simple in *Cheneosaurus* but quite complex in the more conspicuously crested forms. In *Corythosaurus* it runs back from the nose, kinks forward into the crest and there expands into a hollow chamber from the lower part of which a tube runs down to the throat. In *Parasaurolophus* a narrow tube runs right up the upper part of the crest to the top, doubles back and follows the lower part down to the skull and through to the throat. Air drawn in through the nostrils must, of course, pass through the hollow in the crest.

The way of life of the hadrosaurs has been the subject of even more dispute than that of the sauropods. In the early days of research on them one thing seemed obvious: their flattened tails were clearly an adaptation for swimming, and since they had duck-like bills they dabbled in water for their food like ducks, pulling up water-plants and chewing them with the cheek-teeth. The fact that these were clearly adapted for dealing with food far tougher than soft water-plants was disregarded. The truth about their diet was revealed as early as 1922, but, like many discoveries that conflict with orthodox opinion, did not receive recognition at the time. This had to wait until the mid-1960s, when J.H.Ostrom reviewed the subject critically. In 1922 the Senckenberg Museum's mummified *Anatosaurus* was 'dissected' and a mass of fossilized chewed plant remains was found in the region of its stomach. These consisted of needles of a coniferous tree, and twigs, seeds and fruits from other land plants. No remains of water-plants or shells of water animals were found. Evidently *Anatosaurus* sought its food on land, plucking off twigs and foliage of trees and grinding them up in the beautifully efficient mill of its cheek-teeth.

The other controversial feature of the hadrosaurs is, of course, their fantastic bony crests, and most of the speculation concerning them has centred on the hollow crests and the fact that the animal's breath passed through them on the way to its lungs. At first they were considered in the context of an aquatic animal. They could not be snorkels because the nostrils were not at the top but in their normal position on the snout. For some time it was believed that they served to store air to supplement respiration when the dinosaur was under water. One objection to this was that to draw air into its lungs the animal would have either to admit water through the nostrils to take its place or create a vacuum in the crest. The former would entail risk of the water going too far and getting into the lungs, and the latter seems hardly possible. Furthermore, the capacity of the hollow crest was never more than a small fraction of that of the lungs, so the air store would be of very little use even if it could be inhaled.

The acute sense of smell in mammals depends on the turbinate bones in the nasal passage. These are elaborate scrolls of bone covered with a layer of sensory tissue, whose area is thus greatly increased. With the recognition of the hadrosaurs as land animals it is now supposed that the tubes and chambers inside their crests were also lined with sensory epithelium. This would have the same effect and would add the advantage of a keen sense of smell to that of sight and hearing. They seem to have had no means of defence, and to have depended on detection of an enemy's approach in time to take flight. Something of the kind was needed by animals which coexisted with *Tyrannosaurus*.

This interpretation seems reasonable, but it obviously does not tell the whole story. It takes no account of the solid crests and gives no reason for the fantastic diversity of them among the different genera and species. Hadrosaur fossils are sometimes so abundant as to suggest that

OVERLEAF

Left The little theropod called *Compsognathus* is the smallest dinosaur known, about the size of a large chicken. It was also bird-like in structure, and one theory of the origin of birds is that they were evolved from small dinosaurs of this kind.

Right *Allosaurus* must have been one of the most formidable of all the carnivorous dinosaurs, a big powerful animal but not so large as to have been ponderous and slow. It could successfully attack herbivorous dinosaurs much larger than itself, and is shown here feeding on the carcase of one of the huge sauropods.

these dinosaurs lived in herds, just as many large herbivorous animals do today. If a number of herding species of a particular type of animal inhabit an area at the same time, features are generally developed to enable each species to recognize other individuals of its own kind. The distinctive horn shapes and body patterns of the African antelopes illustrate this, and the diversely shaped crests of the hadrosaurs could well have served the same purpose.

Yet another intriguing possibility has been suggested: the air passages in the hollow crests may have been resonating chambers to enable each species to produce a distinctive loud noise, a roar or bellow of some kind, to serve both as an additional recognition character and as a warning uttered by any member of a herd that detects the approach of a predator. Among modern reptiles the geckos are extremely vociferous and some species have quite distinctive calls.

All these assumptions build up a picture of the hadrosaurs as land animals, roaming in herds, ever alert for the approach of one of the large active predators of the time, with every vigilant eye, ear and nose at the service of the entire herd. Two inconsistent features remain: the flattened oar-like tail and, if the mummified fossils have been correctly interpreted, the webbed feet. These strongly suggest an aptitude for swimming and indicate that the hadrosaurs were amphibious. Running away is all very well if you can easily outdistance your pursuer, but there is nothing to suggest that the hadrosaurs were more active than the medium-sized theropods. If they were not, they would need a refuge; they could take this in the large rivers and lakes of the lowland plains they lived on if they fed within fairly easy distance of their banks and fled in good time, warned by the collective sensory equipment of the herd. The carnivore might pursue them into the water, but they would easily outswim him.

The hadrosaurs are surely the most attractive of all the large dinosaurs. Gentle, wary, abundant and fantastically varied in appearance, they must have given the late Cretaceous landscapes a wonderfully dramatic and animated appearance.

Another small group of Cretaceous ornithopods also had a strange specialization of the skull. These were the pachycephalosaurs or dome-headed dinosaurs, in which the roof of the skull was thickened to an extraordinary degree. The better known of the two genera, *Stegoceras*, was a small dinosaur, man-sized or less with a 2-inch (5 cm) thick cranium. In *Pachycephalosaurus*, three or four times as big, the bony helmet could reach a thickness of 10 inches (25 cm), and the snout and hinder part of the skull were covered with bony knobs and spikes. Apart from the skulls, fossils of these dinosaurs are very rare, and it is thought that they may have lived in hilly country where conditions are always unfavourable for the formation of fossils. Some at least of the isolated skulls that have been found are thought to have been carried down from the uplands by fast-flowing streams, rather as boulders of rock are. Few bony structures would stand up to this sort of transport, but these skulls were almost indestructible.

The only suggestion that has been made to account for this strange specialization is based on the behaviour of modern wild sheep and goats. It is thought that the male pachycephalosaurs fought formal duels for possession of the females, running at each other and banging their heads together. The bony helmet would protect the small dinosaurian brain from injury and one or other of the combatants would sooner or later be knocked off its feet or give up from exhaustion.

The armoured dinosaurs are divided into two suborders, the Stegosauria and the Ankylosauria, but it seems likely that they had a common ancestry dating from about the early Jurassic. In strata of that age in southern England an incomplete skeleton of a small ornithischian dinosaur was found as long ago as 1850. It received the name *Scelidosaurus* and seems to display characters intermediate between the two suborders. Until the recent discoveries in the Triassic of South Africa (*Fabrosaurus, Heterodontosaurus*) it was the oldest ornithischian known. It is indeed a primitive member of the Order, as is shown by the lack of a forward prong on its pubis (p. 47), a feature demonstrated by skilful preparation of a second specimen found in 1955.

This was a small dinosaur, 12 feet (3.7 m) long, quadruped, with the hind legs rather longer and stouter than the fore legs, so that the highest part of the body was over the hips. The head, jaws and teeth were small and there were rows of bony plates in the skin arranged longitudinally from neck to tail. Those of the row along the middle of the back were larger and upstanding, giving the animal a serrated dorsal profile. It presumably derived some protection from this armour-plating, though the body was by no means completely covered by it.

Suborder Stegosauria. *Scelidosaurus* is included in this suborder and so is the East African *Kentrosaurus*, whose remains were found in the same Jurassic strata as the famous *Brachiosaurus* fossils. This stegosaur is larger than *Scelidosaurus* and had more effective-looking defensive armour. On each side of the backbone from neck to mid-body there were pairs of small bony plates, but posteriorly and down to the end of the tail these were replaced by long spikes, and there was an extra spike on each side over the hind leg. Neither the plates nor the spikes were part of the skeleton, but grew out from the skin, the plates being attached by their edges, not embedded in the skin. The spikes, especially those on the muscular, lashing tail, must have been formidable weapons. Here again the head was very small.

The largest member of the suborder is the well-known *Stegosaurus*, not very large as dinosaurs go, about 20 feet (6 m) long and weighing less than 2 tons. It has features similar to *Kentrosaurus* but most of them are exaggerated. The hind legs were very long and strong, raising the body like an arch centered over the hips. Along the back were two rows of huge bony plates, implanted in the skin along their basal edges. Near the end of the tail these are replaced by two pairs of long thick spikes, which appear to have had a horny covering in life. The head was relatively tiny. The usual ornithischian beak was present and there were about twenty small teeth in each jaw with only one set of replacement teeth present at a time, a striking contrast with the great dental batteries of the hadrosaurs. There is no doubt it was a herbivore and it probably fed on soft vegetation.

The brain of *Stegosaurus* was minute, about the size of a walnut; a kitten has a brain as large as this. Also there is a cavity in the sacral region which appears to be an enlargement of the great nerve cord that runs through the spine. This was originally described as a second brain, supplementing the rather inadequate one present in the skull. The idea of a creature that thought with its hind-quarters as well as with its head occasioned much merriment in the American popular press. Of course this was not a brain in the sense that it received stimuli from sense organs and passed on relevant instructions to the rest of the body. It is sometimes interpreted as a very large ganglion controlling the massive muscles of the hind legs and tail, but another theory has it that the cavity was not filled with nervous tissue at all, but contained a gland secreting glycogen, a sort of sugar which is readily converted into energy. This would be available as a booster when flight or self-defence called for special exertion. The ostrich has a similar glycogen gland in the same part of the body, and the cavity, whatever its nature, is present in other dinosaurs as well.

The most conspicuous and curious external features of *Stegosaurus* were the great bony plates along each side of the back and the spikes on the tail. The latter are easily accounted for; obviously they were defensive weapons wielded by the big muscular tail and probably very effective. The dorsal plates have always puzzled palaeontologists. It has been assumed that they were a warning signal of some dangerous feature of the dinosaur, like the hood of a cobra. Of what feature has never been clear; the spikes on the tail were plainly visible and warning enough in themselves. The plates were hollow and a recent suggestion is that they were permeated by blood-vessels and served to dissipate excess heat, like the 'sail' of *Spinosaurus*. Another opinion holds that the plates did not stick up vertically but flopped over on each side and covered the back and sides with a sort of cloak of bony armour. I have chosen to illustrate *Stegosaurus* with its more familiar image, with the plates projecting vertically.

Although the stegosaurs had a wide distribution in North America, Eurasia and Africa, they were among the less successful dinosaurs and apparently died out in early Cretaceous times.

Suborder Ankylosauria. The ankylosaurs are the true armoured dinosaurs. They appeared at the beginning of the Cretaceous period, taking the place, ecologically, of the stegosaurs, which disappear from the fossil record at that time. We have no trace of the ankylosaurs' evolutionary history during the Jurassic, but they could well be descendants of a scelidosaurus-like ancestor and could have existed through Jurassic times in an environment not conducive to fossilization.

Acanthopholis is a Lower Cretaceous ankylosaur from southern England and can be regarded as a primitive member of the group, not unlike *Scelidosaurus* in general aspect, with rows of bony plates along its back and sides. *Polacanthus* is another early Cretaceous form with a much better developed defensive armament. Over the neck, shoulders and fore-body there were two rows of strong vertical spikes and a flat shield of bone protected the pelvic region. Behind this two rows of stegosaur-like triangular plates adorn the tail.

In the late Cretaceous the ankylosaurs became very diverse and some of them were large, heavy animals with broad, bulky bodies weighing 3 or 4 tons. Their heads and backs were protected by closely set impenetrable bony plates and some of them had a massive bony club on the end of the tail. In *Ankylosaurus* this was a simple 'blunt instrument', but the tail-club of *Scolosaurus* bore two thick strong spikes. *Palaeoscincus* had a series of large spikes projecting horizontally all round its body close to the ground.

The ankylosaurs were never armoured underneath, and their defence against the prowling biped theropods was probably to crouch down and hope to defy the predator's efforts to turn

OVERLEAF
Spinosaurus, on the left, was a large carnivorous dinosaur which inhabited the region that is now North Africa. *Ouranosaurs* was a herbivore belonging to the other Order of dinosaurs, that lived at the same time and place. Both had a crest or 'sail' on the back, and it is thought that these crests served to dissipate excess heat in an unusually hot climate.

The fossilized bony plates on the back of *Nodosaurus*. This specimen, which measures about twelve feet, gives a clear impression of the texture of this armoured dinosaur's skin.

them over, as they would be helpless and wholly vulnerable if inverted. In forms like *Ankylosaurus* and *Scolosaurus* the enemy's efforts to tip up the 3- or 4-ton body would be embarrassed by heavy crippling blows from the mace-like tail. The big spikes surrounding *Palaeoscincus* were probably not part of the defensive armour but simply gave the animal a wider base and so made it harder to roll it over on to its back.

The teeth were small to the point of being degenerate, but the modern tortoises, which bear some resemblance to ankylosaurs, get along without any teeth at all, biting off their food with the sharp edges of their jaws; doubtless ankylosaurs were equipped with a standard ornithischian beak and fed in the same way.

Convergent evolution produces remarkable similarities in wholly unrelated animals. Until quite recent times there existed in South America gigantic armadillos called glyptodonts; they may have been exterminated by the first men to arrive in the continent. They were large, heavy beasts and had the body covered by a great bony carapace or shell and the skull similarly protected. They were mammalian counterparts of the ankylosaurs, evolved 50 or 60 million years after the dinosaurs became extinct. Some of the glyptodonts had a spiked knob of bone on the end of the tail. This is evidently a useful accessory to the self-defence of a large, heavy, armour-plated animal, and natural selection has produced it in these two instances so widely separated in time and type.

Suborder Ceratopsia. *Triceratops*, with its enormous bony frill, array of forward-pointing horns and general aspect of being some sort of dinosaurian buffalo, is one of the most familiar dinosaur restorations. The representations are probably lifelike as the animal is known from numerous well-preserved fossils. *Triceratops* is the heaviest of all known ornithischians, with an estimated weight of 8½ tons. Only the large sauropods were heavier.

This and the other large horned dinosaurs or ceratopsians are so unlike other dinosaurs that we are fortunate in having fossils representing a stage in their evolution that derives them without much doubt from the ornithopods. In the Lower Cretaceous of Mongolia and China remains have been found of a small dinosaur named *Psittacosaurus* or 'parrot reptile'. It was only 5 or 6 feet (1.7 m) long, with the body, tail and hind legs developed for biped walking, as in *Iguanodon* and other ornithopods, but the fore legs were rather longer and stronger than in most of them; it could certainly walk as a quadruped as well. But its head was high and narrow with a parrot-like beak, the nostril and eye high up on the sides of the head and a ridge of bone across the back of the skull. These are all features that foreshadow the condition in the Ceratopsia. On balance *Psittacosaurus* is regarded as a rather peculiar ornithopod.

In the famous *Protoceratops* we have a dinosaur which is undoubtedly a small primitive type of ceratopsian. It is famous because a fossil breeding-ground of late Cretaceous age was found in Mongolia, with eggs in their 'nests' and the animals in all stages of development, and even traces of embryo bone in unhatched eggs. *Protoceratops* was 5 to 7 feet (2 m) long and was definitely down on all fours, probably because of the weight of its large head, a characteristic ceratopsian feature. The jaws were toothless in front and both upper and lower jaws had a strong beak. The cheek-teeth were numerous and adapted for shearing rather than grinding, a feature carried to extreme lengths in the big fully evolved ceratopsians. In *Protoceratops* the back of the skull grew out over the neck into a collar or shield which was absent in the newly hatched young and developed as the dinosaur grew up. This feature is also present in various exaggerated forms in the large species. The collar was once thought to be a defensive shield; it may have been effective as such in the big ceratopsians, but its primary purpose in all of them was to serve as an attachment for the neck muscles behind and the jaw muscles in front. The head was heavy and the bite very powerful, and both these features resulted in the development of huge muscles needing unusually wide attachments.

After *Protoceratops*, *Triceratops* is the best-known ceratopsian and it is also the largest, with a length up to 24 feet (7.3 m) and a weight twice that of a well-grown elephant. It lived in great numbers in North America in the latest stages of the Cretaceous and must have been one of the dinosaurs that succumbed to the mysterious disaster that exterminated all of them. *Triceratops* had an enormous horn over each eye and a shorter one on the nose, and its bony neck-frill, scalloped at the edge, reached back over its shoulders. We are sometimes puzzled as to how the herbivorous dinosaurs protected themselves against predators, but *Triceratops* leaves us in no doubt. The battery of horns on the huge bony head must have been impregnable so long as their owner stood facing the foe. The fossils are found in some places in such numbers as to suggest that these dinosaurs roamed in herds. If such a herd closed its ranks and faced outward its members would have little to fear even from a hungry *Tyrannosaurus*. Fossils of horns that had been broken and subsequently healed suggest that they were indeed used as weapons.

The cheek-teeth of *Triceratops* are as remarkable as those of the hadrosaurs, but quite different in operation. There was the same system of replacement of worn teeth as these dropped out, but the teeth did not work like a grinding mill. Instead they were sharp-edged and closing of the jaws caused the lower teeth on each side to slide inside and close past the upper teeth like the blades of a pair of shears. The huge muscles that closed the jaws, extending from the neck-frill to the cheeks, were 3 or 4 feet long (1–1.2 m). The food was probably fibrous fronds of palms and cycads, both of which were common in Upper Cretaceous times. The great parrot beak would pluck the fronds out and even the toughest of them would quickly be chopped up into small pieces and swallowed. If the tree was too tall the animal would just push it over with its heavy armoured head. All the ceratopsians, including *Protoceratops*, had teeth and jaw muscles of this kind.

Fossils of ceratopsians have been found only in western North America and north-eastern Asia, and they were quite confined to the Cretaceous period, towards the end of which they evolved explosively. Each genus is distinguished by a different combination of neck-frill and horns. Two main evolutionary lines are distinguished, the long-frilled and short-frilled types. In the former the frill grew over the back; in *Torosaurus*, which was nearly as big as *Triceratops* and had similar horns, the entire skull is 9 feet (nearly 2.7 m) long. *Pentaceratops* had an extra pair of horns growing sideways from the cheek region, making five in all. Among the short-frilled ceratopsians *Monoclonius* had a single horn on its nose, and must have looked rather like a rhinoceros. In *Styracosaurus* the hinder edge of the frill bore a series of large backward-pointing spikes. *Triceratops* belongs to the short-frilled group.

OVERLEAF
The huge sauropods probably fed on the land and rested for a part of each day in shallow water. This scene shows a group of *Diplodicus* entering the water at the margin of a lake in Jurassic times.

The skeleton of *Triceratops*. With a weight of 8½ tons this is the heaviest dinosaur known to us, apart from the sauropods. It was a herbivore with immensely powerful jaws and teeth.

The reason for this great diversity in head armament and ornament may well be the same as in the case of the hadrosaurs, the provison of recognition features in herding animals.

From this brief survey of dinosaur types we will turn now to some general aspects of their natural history. The periods during which the various kinds existed have been mentioned in most cases, but a summary of their history through the long ages of their world-dominance will be useful.

Their story starts in the late stages of the Triassic period, about 200 million years ago, when the small theropods called coelurosaurs, such as *Coelophysis*, became sufficiently distinct from their thecodont ancestors to be called dinosaurs. These were the animals which defeated and rapidly exterminated the therapsids or mammal-like reptiles and drove the small emergent mammals into ecological exile. About the same time the other group of saurischians, the sauropods, appeared on the scene. *Melanosaurus* from the Upper Triassic of southern Africa is the earliest undoubted sauropod known. The prosauropods such as *Plateosaurus* existed at the same time.

Most of the Triassic saurischian types are widespread over the continents, but fossils of ornithischians of this period are rare and known at present only from southern Africa; two genera were mentioned earlier in this chapter, *Fabrosaurus* and *Heterodontosaurus*.

Our knowledge of dinosaur history during the Mesozoic era obviously depends wholly on the availability of fossils. During the early and middle parts of the Jurassic period strata containing remains of dinosaurs are scarce. In southern England a few have been found, including *Scelidosaurus*, *Megalosaurus* and *Cetiosaurus*, representing the stegosaurs, theropods and sauropods respectively. Evidently the dinosaurs were flourishing and evolving during this time, and it is only the record of them that is deficient.

Upper Jurassic strata, dating from about 150 to 140 million years ago, have yielded a great wealth of dinosaur remains. The three most important fossiliferous areas are western North America, in the famous Morrison formation; eastern Africa, in the Tendaguru strata; and several localities in Europe, including Portugal, southern England and Bavaria in Germany. At this time the sauropods were at the peak of their development. Nearly all the well-known genera are Upper Jurassic, including *Diplodocus, Apatosaurus* (or *Brontosaurus*) and *Brachiosaurus*. Many large and formidable theropods are Upper Jurassic in age, including *Allosaurus* and *Ceratosaurus*, and the small active coelurosaurs were numerous and varied. Some wonderfully preserved specimens of these have been found in the Lithographic Limestone of Bavaria, including *Compsognathus*, the smallest known dinosaur. The earliest bird, *Archaeopteryx*, which is now regarded by some palaeontologists as a small feathered coelurosaur, is also known from fossils in this limestone. Among the Ornithischia the stegosaurs were conspicuous and the biped ornithopod *Camptosaurus* is of this age.

In the Lower Cretaceous, 130 million years ago, the sauropods had begun to decline and the theropods were rather like those of the Jurassic. *Megalosaurus*, a very long-lived genus, still existed, and the small but formidable *Deinonychus* was a theropod confined, so far as we know, to the Lower Cretaceous. Among the ornithopods the famous *Iguanodon* was of this age and also the small agile *Hypsilophodon* and the curious 'parrot dinosaur', *Psittacosaurus*, a possible ancestor of the ceratopsians. The ankylosaurs, including *Polacanthus*, were already established and the stegosaurs were on their way out.

The Upper Cretaceous comprises the last 30 or 40 million years of the Mesozoic era, which ended 65 million years ago. During this time the ornithischians evolved into a fantastic array of

OVERLEAF
It seems likely that the huge sauropods spent at least some of their time on land, browsing on the foliage of trees. *Brachiosaurus* could raise its head to a height of forty feet and is seen here feeding from branches far out of reach of other herbivores.

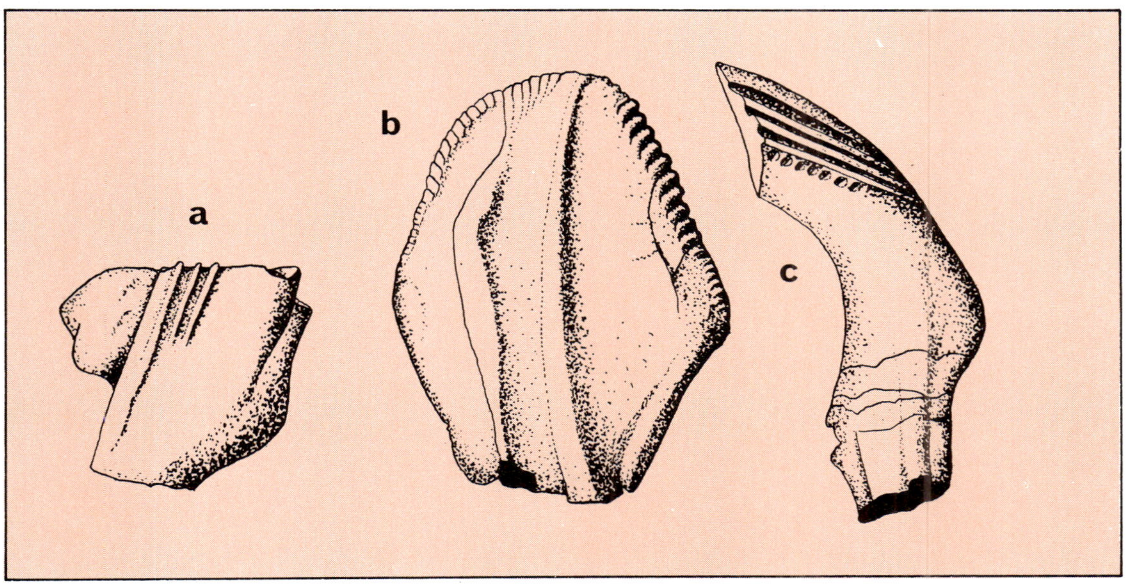

Left Teeth of *Iguanodon*. The worn-down tooth on the left is one of those found in 1822 by Mrs Mantell.

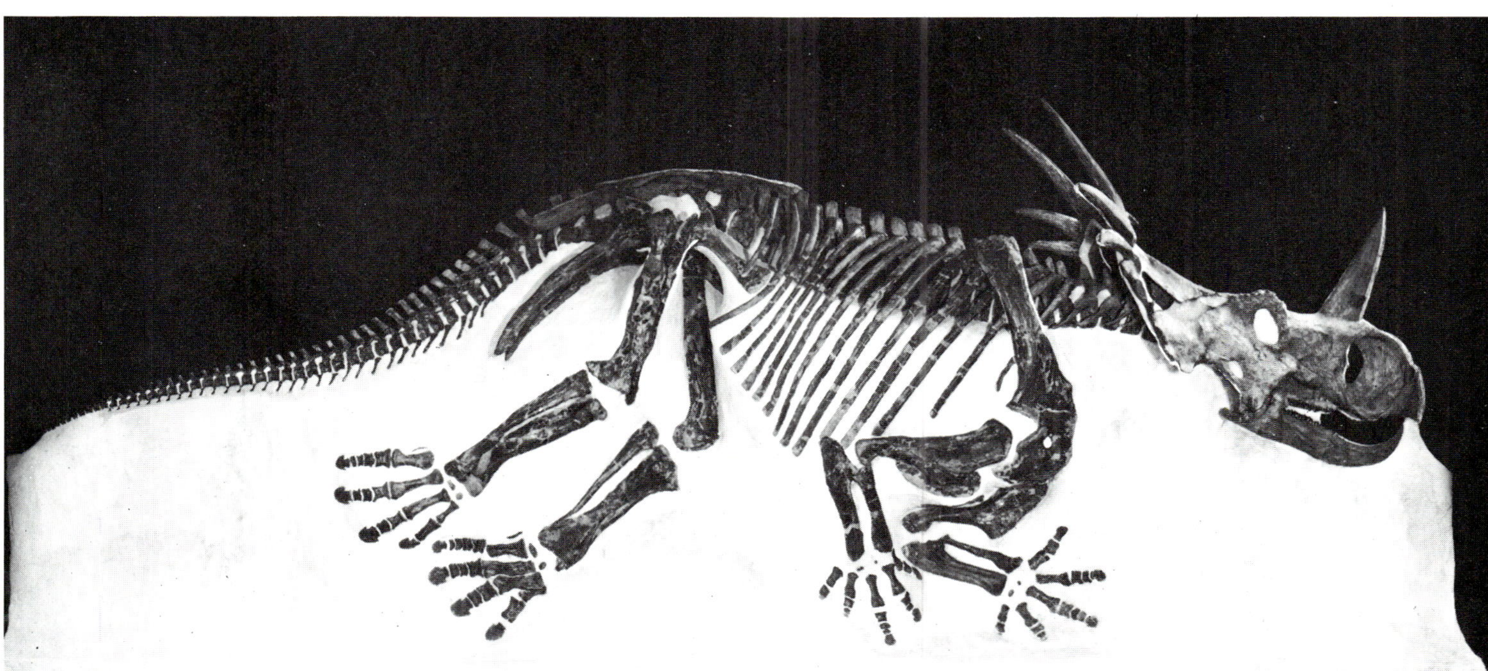

Below The skeleton of *Styracosaurus* a ceratopsian dinosaur. The skeleton is about eighteen feet long.

herbivorous dinosaurs: the alert amphibious hadrosaurs with their fancy head-dresses, the formidable horned ceratopsians, the strange bone-headed pachycephalosaurs and the variously armoured ankylosaurs. The theropods evolved in several interesting directions. The carnosaur line produced the gigantic puny-armed predators such as *Tyrannosaurus* and *Gorgosaurus*. From coelurosaur ancestors were evolved the swift omnivorous ostrich dinosaurs, *Ornithomimus* and *Struthiomimus*, and the small large-eyed, large-brained dromaeosaurs like the Canadian *Stenonychosaurus*. Some of the giant sauropods such as *Alamosaurus* and *Hypselosaurus* persisted right up to the end of the period.

At the end of the Cretaceous there were dinosaurs of almost all the main types known to have existed, herbivorous and carnivorous, giant, medium-sized and small, seemingly well adjusted to their environment, under no threat from any other group of animals. And yet no bone, footprint or any other trace of a dinosaur has been found in any stratum of post-Cretaceous age.

I described the geographical distribution of the Triassic reptile *Lystrosaurus* in relation to continental drift. That of the dinosaurs also reflects world geography before and at the commencement of the break-up of the supercontinents Laurasia and Gondwanaland. Two of the

The skeleton of *Camptosaurus*, a rather smaller relative of *Iguanodon*. It is known from Europe, but most of the fossil material is from North America.

richest sources of Upper Jurassic dinosaurs are the Morrison strata of North America and the Tendaguru formation of Tanzania. Not only are the two faunas extremely rich, they are remarkably similar, comprising dinosaurs of the same types and even of the same genera. Coelurosaurs, sauropods, stegosaurs and hypsilophodon-like ornithopods have been found in both localities, and also the two sauropods *Brachiosaurus* and *Barosaurus*. Here is evidence that the world's dinosaur fauna showed little sign of differentiation in Laurasia and Gondwanaland and was probably rather uniform over all climatically suitable regions of the earth.

In the late Cretaceous, near the end of dinosaur time, interesting new features appear on the map of their distribution. The types with a long history behind them, such as the sauropods, are still uniformly widespread, but the groups that evolved during the Cretaceous have a more restricted distribution. Of these the most conspicuous are the horned ceratopsians, the duck-billed hadrosaurs and the giant carnosaurs or tyrannosaurids. There is some doubt about the interpretation of some southern hemisphere fossils, but, apart from these, members of these groups seem to have been confined to Laurasia. Within the northern supercontinent there was a curious subdivision of faunas in the late Cretaceous. Those parts of Laurasia that are now located in western North America and central Asia had remarkably similar dinosaurs, and the less rich

OVERLEAF
In Belgium a number of fossils of *Iguanodon* have been found in circumstances that suggest that they fell into a deep ravine and were killed. It is possible that they were stampeding in fright from a large carnivorous theropod, and this picture reconstructs that scene.

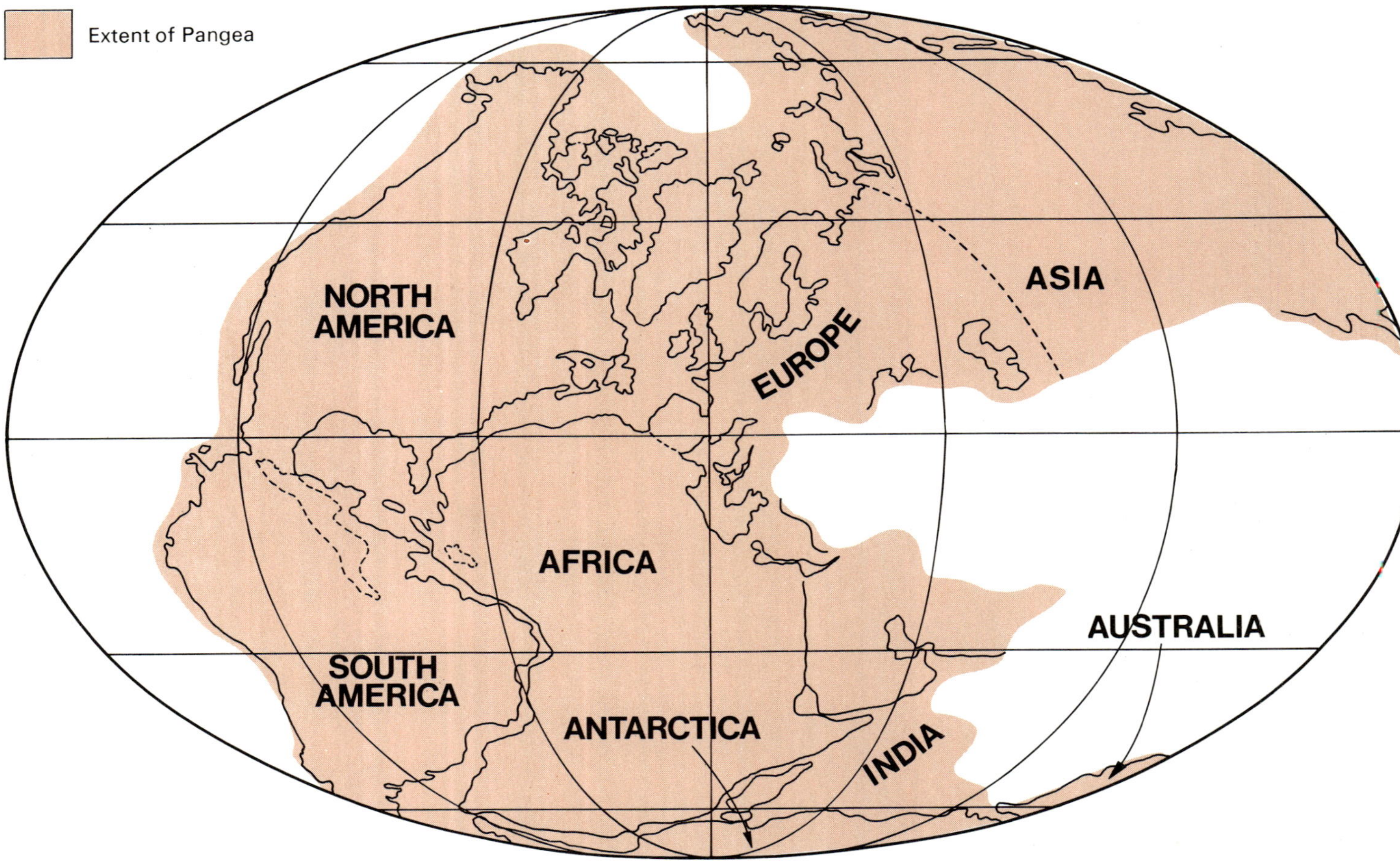

World geography as it may have been in the late Triassic Period, before the start of continental drift. The present continents are shown in outline and the shading shows probable extent of Pangea.

and varied fossil faunas of eastern North America and Europe are also similar.

The most important inference is that by Cretaceous times the Tethys ocean had separated the supercontinents. The other is that seaways, probably extending from the Tethys, somehow formed barriers between the dissimilar faunal areas of Laurasia and allowed migration between the similar ones. It must be remembered that the northern Atlantic rift had not yet completely opened so that North America and Eurasia were mutually accessible. It is clear from the distribution of these dinosaurs that by the end of the Mesozoic the drift which was to break up the supercontinents was already affecting world geography.

Reproduction and Growth

We know tantalizingly little about reproduction and growth to maturity in dinosaurs. Until 1922 no dinosaur eggs had been recognized for what they were, although fragments had been found half a century earlier in southern France and tentatively attributed to large birds. The 1922 discovery was, of course, that of *Protoceratops* in Mongolia, which I have already described in the section on horned dinosaurs. Following this discovery, attention was again given to the find in France, and in 1930 a complete fossil egg was turned up by a farmer ploughing his vineyard in Provence. The search that followed this find produced quite a large number of eggs, all from late Cretaceous deposits. Nine different kinds of eggs have been distinguished, and the largest are probably those of a medium-sized sauropod, *Hypselosaurus*, that grew to about 35 feet (10.6 m), and whose fossil bones occur in the same Cretaceous strata. The eggs are nearly round with a long diameter of about 10 inches (25 cm) and an average content of 3.3 litres. They are rough on the outside with a surface of small lumps or tubercles in place of the wrinkles and ridges on the *Protoceratops* eggs. Some of them were found in groups, usually of five, and it may well be that these are fossil nests, dug in moist riverside sand some 65 million years ago.

Dinosaur eggs of Cretaceous age have been found in Brazil and Montana in the USA, but only in quantity in Mongolia and France. A few Jurassic eggs are known; one from Portugal, which is associated with the bones of a stegosaur, is oval and rather larger than the *Protoceratops* eggs. Two eggs were discovered in the Upper Jurassic Tendaguru strata of Tanzania, but there is no

clue as to which of the numerous Tendaguru dinosaurs laid them.

The *Hypselosaurus* eggs are twice the size of an ostrich's egg, but less than half that of the 8-litre egg of the Madagascar elephant bird *Aepyornis* which was exterminated by man in historical times. This is probably close to the maximum size possible for an egg. In a larger one the gravitional pressure of the internal fluid would cause the egg to break and collapse unless the shell was excessively thick, too thick for the young animal to be able to break its way out. The structure of the dinosaur eggshell appears to be more fragile than that of the large birds, and perhaps that is why the much larger animal laid smaller eggs; possibly the *Hypselosaurus* eggs were at or near the biological limit of size for dinosaurs.

If they are correctly attributed, the most striking thing about these eggs is the enormous disparity in size between the egg, and therefore the newly hatched young, and the adult animal. Let us look at some figures illustrating this theme. In birds the weight ratio of the egg to the parent varies from 4 or 5:1 in the kiwi and some small petrels to about 60:1 in the ostrich. A hatchling kiwi only multiplies its weight four times to become an adult, an ostrich chick sixty times. The large living reptiles lay small eggs for their size, the weight ratio of egg or hatchling to young being 2000:1 in the estuarine crocodile. With 10 tons as a reasonable estimate for the weight of a mature *Hypselosaurus*, it would have to multiply its weight by at least 10,000 in growing up. If its eggs were at or near the limit of size for dinosaurs, then hatchlings of *Brachiosaurus*, with a weight at maturity of 80 tons, would multiply their weight between 50,000 and 100,000 times in growing to full size. These figures are so excessive that some authorities have suggested that the sauropods must have given birth to much larger living young. There is really no evidence of this and the attribution of the French dinosaur eggs to *Hypselosaurus* seems fairly good.

It is a curious fact of palaeontology that young and even half-grown dinosaurs are extremely rare as fossils. We have young *Protoceratops*, of course, and there is a celebrated young sauropod from Utah of the genus *Camarasaurus*. It is a wonderfully complete skeleton of an animal about 16 feet (5 m) long, whereas the adults were up to 60 feet (18 m). The young one has a relatively larger head and shorter neck than the adult.

Fossilized eggs of *Protoceratops*, photographed as they were found in Cretaceous strata in Mongolia.

OVERLEAF
The hadrosaurs were alert and active animals. These two *Anatosaurus* have detected the approach of a great carnivorous theropod, *Tyrannosaurus* in plenty of time to escape into the water.

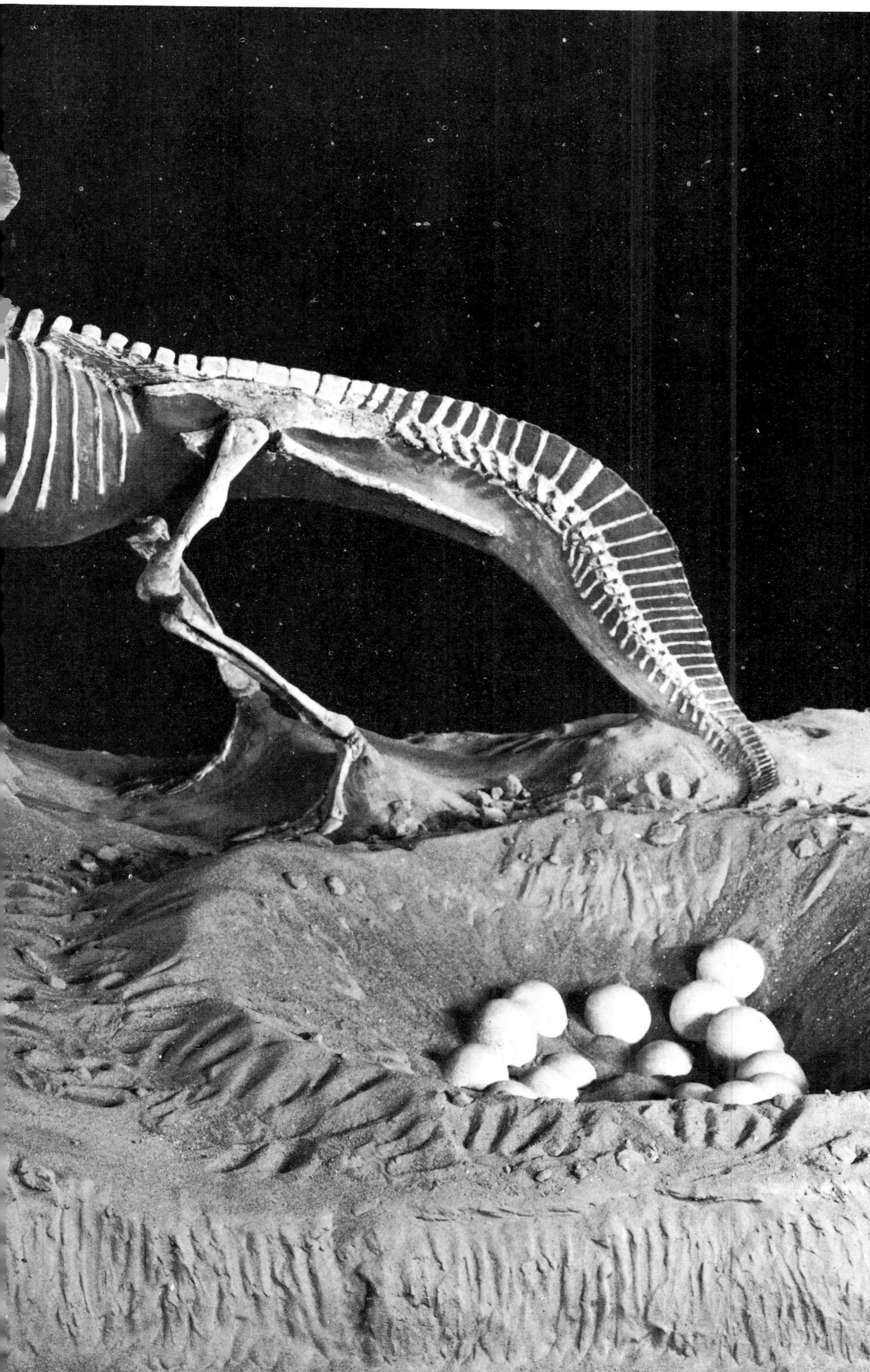

Two *Protoceratops* skeletons with a reconstruction of a nest containing eggs.

OVERLEAF
The crested hadrosaur *Parasaurolophus* lived near the banks of large rivers, feeding on the foliage of trees. The extraordinary crest on its head contained nasal passages believed to have assisted its sense of smell. The gigantic pterosaur *Quetzalcoatlus* is seen soaring overhead.

The big herbivorous dinosaurs such as *Iguanodon*, the duckbills and the ceratopsians are believed to have lived in herds. If they did, there is no evidence that young and adult ran together, nor is there any good reason to suppose that parental care was a feature of dinosaur life. Most likely the young dinosaurs inhabited an environment well away from that of their huge parents and the great predators that lived with and on them. It may well be that after hatching they migrated to the uplands, only returning to the fertile but dangerous river plains when appetite and size made it expedient and safe for them to do so.

Even if young dinosaurs grew fairly fast the big ones had an enormous amount of growing up to get through before they reached full size. This leads immediately to the question: what sort of age did they live to? Fossil bone from the neck-frill of a mature Upper Cretaceous ceratopsian has been shown to be formed of thin alternating dark- and light-coloured layers totalling about 120 pairs. It is claimed that these may be analogous with the annual rings seen in the wood of trees growing in a seasonal climate, and therefore indicate that the dinosaur was about 120 years old when it died. There could be wide variations on each side of this single sample. The small coelurosaurs probably had a life-span comparable with our own; *Brachiosaurus* might well have lived for two centuries or more.

The skull of *Pachycephalosaurus*. The domed skull roof is formed of solid bone.

Chapter Four

Were They Reptiles?

OVERLEAF

The dome-headed dinosaur *Pachycephalosaurus* had a skull with a ten-inch thickness of bone in front. It is possible that the males fought duels, just as modern wild sheep and goats do, by charging and butting with their heads.

A.J. Riley

Ever since dinosaurs became a subject of general knowledge they have been treated with a curious mixture of wonder and amused contempt. The wonder is inspired mainly by their great size; contempt by the conception of them as huge, cold-blooded, slow-moving, puny-brained reptiles, doomed to extinction by their total inferiority to the mammals which superseded them. This has been the popular view. Informed opinion, held among palaeontologists, has been less extreme. It is accepted that they cannot have been inefficient animals since they dominated the land for 120 million years. It has also been realized for a long time that the mammals came into their own due to some environmental hazard that exterminated the dinosaurs, not to any sort of competition in which the mammals emerged victorious. The picture of the cold-blooded, slow-moving dinosaur, however, has been generally accepted, at any rate for the large ones. The justification of this view is that animals with a reptilian metabolism could not mobilize and maintain the great amount of energy needed for rapid movement of bodies weighing many tons. The smaller bipeds, like the ostrich dinosaurs, have always been given credit for being fairly active and speedy, but with a 'cold-blooded' metabolism could not have sustained rapid movement for long. The notion that the dinosaurs were reptiles and nothing more has never been questioned until recent years.

The first dissenting voice was that of John Ostrom in 1969, who argued that dinosaurs were warm-blooded and therefore not reliable indicators of a hot climate. Their champion, however, is his pupil Robert Bakker, who brought the sauropods out of the lakes and swamps on to the forested dry land. He maintains that the dinosaurs were not only warm-blooded, but were just as active and speedy as the modern large mammals. The terms 'cold-blooded' and 'warm-blooded' do not accurately describe the conditions they refer to, and I will lose no time in substituting for them the more precise expressions 'ectothermic' or 'externally heated' and 'endothermic' or 'internally heated', and in trying to explain what they really mean.

A lizard is an ectothermic or 'cold-blooded' animal. Many lizards spend their active lives running and basking in sunshine. By a pattern of behaviour that exploits and avoids the sun's rays at need they maintain their internal temperature at a constant level that may be higher than our own 'normal' condition. The greater earless lizard of the American south-west keeps its body within 3.3°F (1.8°C) of its average temperature of 101.3°F (38.5°C) during the greater part of its active life. At night, or if it is kept in the shade as most captive lizards are, its temperature falls and it becomes completely or relatively inactive, but suffers no harm unless the fall is drastic. During its natural daily life it is rather more warm-blooded than we are ourselves. Another American species, the whip-tailed lizard, stabilizes its daytime temperature at about 105°F (40.5°C), a level indicating high fever in a man.

Man, and most mammals and birds, maintain a constant internal temperature independently of external conditions. This is achieved by a high and accurately controlled rate of metabolism, oxidizing or 'burning' carbohydrates and fats to keep their insides at a level of warmth usually rather higher than their surroundings. External cold is counteracted by more rapid consumption of fuel, and in conditions of excessive heat the temperature is kept down by panting or sweating, which evaporates water. If their internal temperature falls much below the specific constant, usually within a few degrees of 100.4°F (38°C) in endothermic animals, death from hypothermia ensues. A normal accompaniment of endothermy is an insulating cover of some kind, fur in mammals, feathers in birds, and in ourselves clothing to suit warm or cold external conditions. The effect of the cover is always to trap or hold air, which is a bad conductor of heat. The endothermic condition, with its unvarying optimum temperature, allows for the maintenance of a more highly organized and sensitive nervous system than the ectothermic.

The question of endothermy and ectothermy is fundamental to the recent new look at the dinosaurs, but before considering them in the light of it, I have to make a digression on the subject of absolute size in bodies of any kind. There is a mathematical relationship which constantly affects our experience, but of which by no means everybody is aware in exact terms. It states that if two bodies are of the same substance and shape, but of different sizes, the smaller one will have a larger surface area relative to its weight or mass. A one-inch cube of wood weighing one ounce will have a surface area of six square inches. A six-inch cube of the same wood will weigh $6 \times 6 \times 6 = 216$ ounces, and its surface area, 36×6, will be 216 square inches. The smaller one has an area of six square inches to the ounce, the larger one of one square inch to the ounce. A pound of small potatoes takes longer to peel than a pound of large ones because the small ones have a greater total area of peel. A roll has a much larger proportion of crust to crumb than a loaf.

The relevance of this to body temperature in animals is that since heat enters and escapes through the surface a small animal, with its large relative surface area, will heat up in the sun and cool down in the shade more quickly than a large one. As we have seen, a lizard can keep a high constant temperature by simply dodging in and out of the sunlight, because its temperature response to the presence or absence of radiant heat is almost immediate. A ten-ton dinosaur would lose and gain heat to and from its surroundings so slowly that even the alternation of night and day would have a negligible effect on its body temperature. Such an animal must have maintained its internal heat at an optimum and unvarying level by the action of its own metabolism; it was in fact endothermic and in a warm climate was more likely to suffer from overheating, under stress of exertion, than from chill due to weather conditions. It is now generally accepted that this is how the big dinosaurs lived, but the question then arises: is endothermy of this kind necessarily accompanied by a high metabolic rate?

'Yes', say Bakker and his supporters.

'No', say his critics.

I will not presume to pass judgment, but will try to summarize the opposing arguments.

The advocates for a high metabolic rate for dinosaurs point out that they had a fully erect limb posture, in which the body is held well above the ground when the animal is moving, and quite long limbs. These features indicate active movement which, in such large animals, would require an enormous amount of energy and therefore a high rate of metabolism. Bakker suggests that dinosaurs may have been able to run at speeds as high as those attained by the modern horses and antelopes. Support comes from a totally different field of investigation, that of the internal microstructure of the bones. This shows that dinosaur bone was abundantly supplied with blood-vessels and with Haversian canals, which control the exchange of calcium between the skeleton and the blood. In this respect dinosaur bone resembles that of mammals but not that of the modern reptiles. The possession of a complete secondary palate, which allows an animal to breathe while it is eating, is also a feature of dinosaurs; a high metabolic rate requires continuous breathing. In some of the largest dinosaurs there are great cavities in the vertebrae which may indicate an air-sac system, like that described for birds and attributed to the pterosaurs.

Bakker also advanced an argument based on 'community structure', maintaining that a low numerical proportion of predators to prey animals indicates that the predators must have a vigorous metabolism, and that some fossil dinosaur communities are of this kind. This notion is so involved, and has been so effectively criticized, that I shall not elaborate it further.

Opponents of Bakker's views have objected that a high metabolic rate in big dinosaurs would require such vast quantities of energy that they would never be able to find, consume and digest the food needed to supply it, even if they ate continuously for twenty-four hours a day. It is also rather difficult to reconcile a high rate of metabolism with the exceedingly small brains possessed by all the large dinosaurs. It is harder still to imagine them running at high speeds, since such athletic behaviour in an animal weighing several tons would require nervous coordination of a very high order. This notion of large dinosaurs racing and galloping about over the Mesozoic landscapes is indeed regarded with rather general scepticism.

The emphasis in this discussion has been up to now on the large dinosaurs, which were endothermic because they could not help it, and which may or may not have had a high rate of metabolism as well. With the small ones, and also the young of the large kinds, the case is not quite the same; their size did not make obligatory endotherms of them, and they could have lived as ectothermic reptiles, just as the crocodiles and large lizards do today. It has been argued that the young of large dinosaurs were likely to have been endothermic because they would have to have grown very fast and this would require a high metabolic rate. On the other hand, lacking any sort of insulating cover, they would have been hard put to it to maintain a high body temperature at night or during spells of cool weather. It seems more likely that the hatchlings were not endothermic, but acquired this feature when they attained the bulk needed to maintain their internal heat at a constant level.

The case of the dinosaurs that never grew to a large size is different again. Most of the coelurosaurs were a bit larger or smaller than a man, some a good deal smaller, and there were ornithischians, such as *Hypsilophodon* and the primitive Triassic *Fabrosaurus*, which were also quite small. One authority, L.B.Halstead, thinks *Fabrosaurus* was an ectotherm; as a small primitive dinosaur it may well have been so. But the late Cretaceous ostrich dinosaurs such as *Struthiomimus* require separate consideration. This animal was a little larger than an ostrich and, of course, it had a tail, fore limbs not modified as wings, and no feathers. Otherwise the skeletal proportions and posture of the two were quite remarkably similar. Now the ostrich is one of the fastest of all running animals, performing in a way that no ectothermic creature could match. Evolutionary studies show that structural convergence nearly always indicates close similarity

OVERLEAF
Stegosaurus is one of the best known herbivorous dinosaurs. The big spikes on its tail were obviously defensive weapons, but the bony plates on each side of its back remain a puzzle. They may have been part of a warning display of some kind, or they may have served as radiators to dissipate heat.

in habits. *Struthiomimus* would never have been so like an ostrich unless it had the power to run in a similar way. An ectothermic animal could never mobilize enough energy to run at such a speed, so *Struthiomimus* must have been endothermic. So runs the argument of the advocates of high activity and hot blood for small as well as large dinosaurs. *Hypsilophodon* also seems to have been built for speed and activity, and so do the small ferocious dromaeosaurs like *Deinonychus*. Holders of extreme views in favour of endothermy even extend it to the biped thecodont ancestors of the dinosaurs, and to the primitive dinosaurs themselves, such as *Fabrosaurus*.

The Origin of Birds

There is an interesting argument in favour of endothermy being a character of all the dinosaurs: although some were of no great size, no really small ones are known; there seem to have been none of the rat- or mouse-sized dinosaurs that might have been expected to appear at some time in their immensely long evolutionary history. But if they were endothermic and lacked any insulating cover this avenue of evolution, open to the mammals and birds and to the ectothermic lizards, would be closed to them. A tiny endothermic dinosaur could not keep itself warm and would die if it became chilled. The chicken-sized *Compsognathus* seems dangerously small for a naked ectotherm. But was it naked, or was it perhaps covered with feathers? This brings us, of course, to the intensely interesting question of the origin of birds.

There is no doubt that birds evolved from some type of archosaurian reptile, that is to say from a thecodont or one of the branches arising from the thecodont stem, but there is still a lack of agreement about what particular sort of archosaur was their immediate ancestor. This is perhaps rather surprising in view of the fact that we have a series of well-preserved fossils of an animal perfectly intermediate between an archosaurian and a typical bird. This is of course the famous *Archaeopteryx lithographica* from the Upper Jurassic Lithographic Limestone of Bavaria, which has yielded so many remarkably preserved fossils of pterosaurs, insects and other delicately constructed animals. The two earliest discovered specimens of *Archaeopteryx* are generally regarded as the most precious fossils in the world.

The discovery of the five known *Archaeopteryx* fossils makes an interesting story. The first one was found in 1861 and was acquired by Dr Karl Häberlein, who took fossils from the Solenhofen quarrymen in lieu of medical fees. He offered it for sale for £700, and an emissary from the British Museum secured it for this sum, together with a large number of other fossils from Dr Häberlein's collection. The doctor used the money to provide a dowry for his daughter. Very shortly before this discovery a single feather had been found and named *Archaeopteryx lithographica* by H. von Meyer. The name means 'ancient feather from the Lithographic Limestone', and it was transferred to the more complete fossil, which now stands as the type specimen of the species. It has been worked on and 'developed' by the staff of the British Museum and a great deal of information extracted from it. The head of this specimen is fragmentary, but an internal cast of the brain is preserved. Coming as it did so soon after the publication in 1859 of Charles Darwin's *Origin of Species*, it gave a tremendous boost to the supporters of the new and highly controversial theory of evolution.

The second specimen was found in 1877 near Eichstätt, ten miles away from the first find, and was acquired by Dr Häberlein's son and purchased for the equivalent of £1000 for the Berlin Natural History Museum. It is a finer specimen than the first, with the whole skeleton, including the skull, intact. It has not been developed and scrutinized so thoroughly as the one in London, probably from a desire to keep so beautiful a fossil in its pristine condition. The third *Archaeopteryx* came to light in 1956, quite near the site of the first, but at a geological horizon six to eight metres lower. It is less perfect than the other two, lacking the head altogether, and is now on display at the Erlangen Museum near Solenhofen.

In 1970 Dr Ostrom was examining the pterosaur fossils in the Teyler Museum at Haarlem in Holland. One specimen from the Lithographic Limestone of Bavaria struck him as having limbs not conforming to the pterosaur pattern, and in the course of his close examination of it he found faint impressions of feathers. There is no doubt that it is an *Archaeopteryx*, but it is incomplete and here again the head is missing. It came from the quarries in 1855, so it is really the earliest specimen found, though only the fourth to be recognized. The discovery of the fifth specimen was announced in 1973. It was found in 1951 near Eichstätt and was first believed to be a very small coelurosaurian dinosaur, similar to *Compsognathus* but even smaller; it is indeed smaller than the other specimens of *Archaeopteryx*. Here again oblique lighting has revealed traces of feathers. It is virtually complete but has not yet been fully described.

The first fossils indicate an animal about the size of a crow, which was not so much a blend of

The fossil of *Archaeopteryx* found in 1877 and preserved in the Berlin Museum.

reptile and bird as a patchwork, parts being wholly bird and parts wholly reptile. The head was reptilian with toothed jaws. The brain was larger than is usual in reptiles, but the large cerebral hemispheres and cerebellum found in modern birds were not present. The whole vertebral column is like that of a reptile and includes a long tapering tail, and the hand was formed of three long free fingers. There was no keel on the breastbone for the attachment of large flight muscles and respiratory air-spaces in the bones were not developed.

The main avian skeletal characters are the presence of joined clavicles, forming a wish-bone; an 'opisthopubic' pelvis, with a long backwardly pointing pubis like that of birds and ornithischian dinosaurs; a big toe rotated backwards and opposed to the other toes to form a foot adapted for grasping and perhaps for climbing. Nevertheless there is little doubt that if feathers had not been clearly visible in these two marvellous fossils no one would have regarded *Archaeopteryx* as anything but a reptile. But the feathers are there, exactly like those of modern birds in their microscopic structure and in their mode of attachment to the fore limb. This precise correspondence leaves no doubt that *Archaeopteryx* has an ancestral relationship to all birds, but what were the ancestors of *Archaeopteryx*?

As I have said, the reptilian forebears of birds were archosaurs just as surely as those of the mammals were therapsids. Although the resemblance of *Archaeopteryx* to a small coelurosaur was noticed long ago, no one dared to suggest that birds were descended from dinosaurs. The orthodox view until three or four years ago was that the birds branched off from the thecodont stem, from a Euparkeria-like ancestor, in the Triassic, and so originated separately from the dinosaurs and at about the same time. Against this theory was the complete absence of any fossil suggesting a bird for the next 70 million years, and then, in the Upper Jurassic, an animal that was still half reptile. One would expect the birds to have evolved and established themselves rather faster.

The chief objection to a dinosaur ancestry was the presence in *Archaeopteryx* and all subsequent birds, and in the thecodonts, of clavicles or collar-bones; these were apparently lost

OVERLEAF
Palaeoscincus (right) and *Scolosaurus* are two examples of the armoured dinosaurs or anklyosaurs. The tail of *Scolosaurus* was adapted to provide a formidable spiked club to be lashed about if the animal was attacked. The lateral array of spikes around *Palaeoscincus* may have made it more difficult to turn it over on its back.

by the dinosaurs right at the beginning of their evolution. A rather insecurely founded dictum called Dollo's Law states that once a character is lost by an evolutionary lineage it can never be regained; that evolution cannot go into reverse. *Archaeopteryx* could not be a small feathered dinosaur because it had a wish-bone and dinosaurs had irretrievably lost their clavicles. In 1973 the indefatigable Ostrom found that some coelurosaurs do have clavicles, and suggested that these might have been present throughout these small theropods' history, but were developed only as cartilage and so not preserved. Here we must remember that the recently found Eichstätt *Archaeopteryx* was identified as a coelurosaur in the 1950s, when palaeontology was already a sophisticated science; only detection of the feathers revealed it as a bird.

Although an attempt has been made to link the ancestry of the birds closely with crocodiles, most palaeontologists are now agreed that they originated, probably in Jurassic times, from some of the small theropods known as coelurosaurs, which, being so small, had to grow feathers to keep warm; feathers could well have been derived from scales which became frayed at the edges, and which could have formed the beginnings of an insulating cover before they became true feathers. Using feathers to fly with was an evolutionary afterthought to which I will return.

I expect most readers have already spotted the paradox in this theory of dinosaur-to-bird evolution: the coelurosaurs were saurischians, reptile-hipped dinosaurs with the pubis pointing forwards; in *Archaeopteryx*-and-all-birds it points backwards. Surely if birds have a dinosaurian ancestor it should be among the bird-hipped ornithischians. This has indeed been suggested, but it is not now considered seriously. *Archaeopteryx* has the jaws and teeth of a saurischian and, except for its pelvis, follows the saurischian pattern wherever this differs from the ornithischian. Ostrom claims that in both the London and Berlin fossils the pubis was dislocated after death and displaced so as to point misleadingly backwards. This may be so (it seems rather unlikely in both fossils), but even if it is true, we have modern birds with the pelvis of ornithischian type and their presumed ancestors, the coelurosaurs, with a saurischian pelvis. At some time in the development of the birds, whether before or after the *Archaeopteryx* stage does not really matter, evolution brought about a profound change in the position of the pubis. Why was this necessary, when the coelurosaurs certainly stood, balanced and ran most efficiently on their reptile hips? A bird's hind limbs and pelvis have to serve as landing-gear as well as for locomotion; perhaps this has something to do with it.

I have already referred to the attractive theory that as dinosaurs were endothermic they could only evolve into small animals if an insulating coat of some kind was developed, and that some coelurosaurs of the *Compsognathus* type and size changed their scales, in the course of their evolution, to feathers. The question of how feathers evolved further to provide wings brings us to the question of what was the way of life of *Archaeopteryx*, a matter as hotly debated as that of its ancestry.

All are now agreed that this bird could not really fly. There are no skeletal attachments for strong flight muscles and its brain was not adequate to control and coordinate the intricate movements needed for true flight. It had very much the build of a biped runner, but the hind feet seem adapted for grasping, and the animal could have spread out a large enough flat surface of wing and tail feathers to support it in the sort of gliding performed by a flying squirrel The issue is between these alternatives: did it run on the ground or did it climb among the branches, gliding from time to time from one branch or tree to another?

If it ran on the ground what use were its wings? Even if it could attain speed for a take-off like an aeroplane, it could not remain airborne, and its long feathered tail would be an encumbrance. If we look at some of the modern gliding animals, particularly the flying squirrels and marsupial flying phalangers of Australia, we see a similar combination of supporting surfaces on each side of the body and a long tail behind, which is furred all round in some species, but in others has a flat fringe of hair on each side, functionally very like the bilaterally feathered tail of *Archaeopteryx*. It has been objected that a biped running animal like a coelurosaur would be unlikely to evolve into a climber, but I find this easier to believe than the strange notion, which has been seriously proposed, that the wings of birds first evolved as a sort of butterfly net to catch insects running on the ground. *Archaeopteryx* was most likely an arboreal glider, an obvious and logical stage for birds to pass through in 'learning' to fly.

When the birds became established in the form in which we know them, well before the end of the Mesozoic, they must have quickly displaced the pterosaurs in the forests and in all habitats near the ground. This was mainly due to the unimpaired functioning of their hind limbs for perching, hopping and running. In pterosaurs (as in bats) the hind limbs were involved with the wings. In one realm, that of soaring flight, the pterosaurs remained supreme until the end of the Cretaceous.

Were Dinosaurs Reptiles?

Reverting to the subject of dinosaurs in general, and particularly to the title of this chapter: if we admit that both big and small ones were endothermic, are we justified in regarding them as reptiles? We have agreed that the highly adapted warm-blooded pterosaurs were as distinct from reptiles as birds are. If they are elevated to Class status in taxonomy, should not dinosaurs be regarded in the same way? The followers of Ostrom and Bakker go further. Bats, they say, are winged mammals, but no one suggests that they should be set wholly apart from the mammals on this account. Birds have been shown to be winged dinosaurs, no more entitled to classification on their own than bats are. There should be a taxonomic Class Dinosauria, ranking equally with the Mammalia and Reptilia and divided into three Orders, Saurischia, Ornithischia and Aves or birds. Perhaps some people may be shocked by this, but I like to think that I have dinosaurs singing and nesting in my garden, and that I ate a dinosaur's egg for my breakfast this morning.

Why Did They Die Out?

However, for the sake of convenience I shall continue to use 'dinosaur' in its more widely accepted sense, and conclude this chapter with a brief examination of the greatest puzzle in palaeontology: why did the dinosaurs become extinct and, in particular, why did they die out with dramatic suddenness at the end of the Cretaceous period?

Numerous reasons have been advanced, ranging from the wrath of God to plagues of caterpillars consuming the foliage on which the dinosaurs subsisted. A theory which held the field for a long time was that of racial senescence: they grew old and degenerated as a group because they had existed for so long. The neck-frills of the ceratopsians and the strange nasal horns and crests of the hadrosaurs were regarded as 'over-ossification' due to some hormonal imbalance brought on by racial old age. There are no grounds for any analogy between the individual life-span and the duration of a race or type of animal. The ancestries of all animal groups go continuously back to the primeval ocean where life began; the division of them into groups is an intellectual exercise of our own to assist classification, not a reality in nature. Furthermore, the so-called grotesque features of the late Cretaceous dinosaurs most probably were all well-adjusted adaptations to their way of life. They were flourishing at that time as never before: ponderous sauropods and ceratopsians, armoured ankylosaurs, big alert hadrosaurs, huge predators, swift ostrich dinosaurs, small active big-brained dromaeosaurs – by no means an assemblage to suggest senescence or degeneration.

The great evolutionary change-over in terrestrial plant life, the triumph of the flowering plants, came 35 million years before the extinction of the dinosaurs. So far from contributing to their demise, the onset of the new floras seems to have stimulated their evolution by its offer of a diversified food-supply. Finally, the dinosaurs were the only large terrestrial animals in existence; there was no sort of threat or challenge to them from any other group of animals.

It is important to remember that the dinosaurs were not the only animals to suffer extinction at the end of the Cretaceous. The pterosaurs also failed to survive, and in the surface waters of the sea the plesiosaurs and mosasaurs disappeared with equally dramatic suddenness. No one could impute 'racial senescence' to the mosasaurs, which first appeared after mid-Cretaceous times. For some reason, probably not connected with the terminal Cretaceous extinctions, the ichthyosaurs dwindled and disappeared a little after the middle of the period. Among the invertebrates the ammonites continued to the end of Cretaceous times but no later, and so did the majority of the planktonic foraminifera (minute protozoans with shells) of the chalk, which is the latest marine Cretaceous formation. The ammonites, with their chambered gas-filled shells, are believed to have floated in the open sea, and all the marine animals which suffered sudden extinction appear to have lived in the surface waters.

Turning again to the land, the difficulty encountered by all attempts to explain the disappearance of the dinosaurs is that they should have gone, and the mammals, birds and all types of reptiles that have survived into our times should have been spared. The crocodiles, turtles, lizards and snakes were all coexistent with the dinosaurs at the end of the Cretaceous. Why are all these still with us when the dinosaurs, which seem to have excelled the whole assemblage of modern reptiles in diversity and adaptability, failed to pass some drastic test of survival?

No signs of glaciation have ever been found in the Cretaceous geological record, but it is now widely believed that the agent of extermination was a gradual or sudden world-wide onset of

OVERLEAF
Left Heads of five genera of hadrosaurs to show diversity in the form of the nasal crests. Top row: *Kritosaurus*, *Saurolophus*, centre: *Corythosaurus*, *Lambeosaurus*, lower row: *Parasaurolophus*.

Right *Psittacosaurus* means 'parrot reptile', and this refers to the very parrot-like beak possessed by this small dinosaur from the Cretaceous of central Asia. Although it is classified with the ornithopod dinosaurs it has features which appear to be ancestral to those of the ceratopsians of which *Triceratops* is a well known example.

cold conditions, probably seasonal, affecting all latitudes, cooling the land and the surface waters of the sea and exterminating those animal types least able to resist or adapt themselves to cold. The fact that the surface-water oceanic animals were the marine victims fits this notion well, as only the surface waters are directly warmed by the sun; the deeper, colder waters would be very slowly, if at all, affected by atmospheric chilling. The theory also has the support of the advocates of endothermy for all dinosaurs. The mammals and birds had a warm insulating covering and the ectothermic land reptiles could allow their temperature to fall and go into winter hibernation. Most of them were also small enough to burrow in the earth and so escape predation, and to some extent the cold as well, when they became chilled and inactive. But endothermic naked dinosaurs of whatever size were doomed. Even if the largest kinds were protected from chill by their great size, and were ectothermic when young, they would have suffered from the cold at some stage in their growth. By the end of the Cretaceous the feathered coelurosaurs (if the suggestion that they existed is correct) had evolved too far to carry on any tradition of being saurischian dinosaurs; they survived, but only as birds.

In the previous chapter I mentioned the dinosaurs' eggs found in the south of France. They are in a series of strata covering a short period, geologically speaking, just before and up to the end of the Cretaceous. A detailed analysis has shown that the shells of the eggs in the older layers are quite thick, up to 2.5 mm, but in the higher and younger layers they become progressively thinner, down to 1 mm; such large eggs would be dangerously fragile with such thin shells. Probably the dinosaurs in that region and at that time were undergoing some kind of stress, and this could well have been due to increasing cold.

The pterosaurs are a special case. They had a furry covering, but by that time they had been ousted by the birds from all but the soaring mode of life. It may well have been turbulent weather, accompanying the cold, that put an end to them. If *Quetzalcoatlus* was a carrion feeder it would have starved when all carrion was reduced to the size of a rabbit or less; at the end of the Cretaceous there were probably very few land animals larger than this.

We do not know what caused this world-wide cold, but it was apparently something quite different from the conditions that bring about ice ages. Two suggestions have been made: the one most favoured at present is that an asteroid, a mass of rock some miles in diameter, struck the earth. The colossal impact would have thrown millions of tons of dust up into the atmosphere, which would have greatly reduced the sun's radiation for a period of several or perhaps many years. This would have lowered temperatures over the whole earth and also reduced the rate of plant photosynthesis which is the basis of the food supply of all animals.

A supernova explosion of a star has also been suggested as having a similar effect. If it had been much nearer the earth all life would have been destroyed by its emission of cosmic radiation on a gigantic scale, and remote supernovas, like that of AD 1054, of which the Crab Nebula is the remnant, are merely exciting astronomical events. One at an intermediate distance might have the greater part of its radiation absorbed by the ozone layer and the ionosphere, but this would bring about high-level turbulence. Water vapour displaced into the upper atmosphere would then condense to form a dense and persistent cloud cover.

The most serious obstacle to both these theories is that in terms of geological time they would have been sudden catastrophes. But the fossil record shows a decline of the dinosaurs that was rapid and complete, but progressive. The decline was in diversity as well as abundance; in every group the number of genera decreases as one proceeds upwards in the Upper Cretaceous strata, until the point comes when no dinosaur fossils at all are to be found. The marine ammonites also decrease in variety before they abruptly disappear at the same level in the fossil record. The advocates of catastrophe maintain that this is apparent rather than real, and due to less intensive searching in the highest Cretaceous rocks. But would this be likely to apply to dinosaurs and ammonites as well? We will leave the matter there and hope that more evidence will soon be forthcoming.

It has been said that if the mammals had not been overtaken and eclipsed by the dinosaurs in the Triassic man might have been on the moon in the Cretaceous period; that is to say that if the rapid expansion of the mammals had taken place 120 million years sooner than it did, an intelligent being, at least something like man, would have appeared many millions of years earlier than the time of our own evolutionary emergence.

The development of life on the earth would also have followed a very different course if the dinosaurs had not met with mysterious disaster at the end of the Cretaceous. It is impossible to say how the world would now be populated, but one thing is fairly certain: we should not be here, because the evolutionary radiation of the mammals which led up to man would not have come about.

Chapter Five

Men and Dinosaurs

OVERLEAF

Some of the most spectacular of the dinosaurs lived at a time shortly before they all became extinct. Among these were the greated horned dinosaur *Triceratops* and the remarkable crested hadrosaur *Corythosaurus*.

In places where fossils are abundant and well-preserved they are bound to be noticed, and people have noticed them from earliest times. Some of the Greek philosophers, including Herodotus (*c.* 485–425 BC), had sensible if rather imprecise ideas of their real nature, and the great genius Leonardo da Vinci (1452–1519) thought so clearly about them that he could have laid the foundations of geological science. He was, however, a prudent as well as a brilliant man and recorded his dangerous ideas in secret notebooks which have only come to light quite recently. The orthodox medieval scholars held beliefs about fossils that seem very strange today, but any sort of clear thinking about them would have conflicted with the biblical story of the creation, and its expounder would have been in serious trouble. They invoked a 'plastic force' or 'formative virtue' that produced fossils out of the earth; one sixteenth-century writer believed that fossil shells were generated by 'the tumultuous movements of terrestrial exhalations', but no one cared, or dared, to question absurdities of this kind.

These ideas were applied mostly to fossil shells and other invertebrates. Large bones presented no problem: they were the remains of men who had perished in the Flood; those of mammoths or dinosaurs were attributed to giants. Enlightenment came slowly in the sixteenth and seventeenth centuries by steps which I have not space to trace in detail. Fossils came to be recognized as remains of animals that had lived, mostly in the sea, in past ages. The presence of marine fossils far from the sea gave colour to the notion of a universal deluge, or a series of catastrophes of this kind, and 'The Flood' continued to bedevil geology until the publication of Charles Lyell's *Principles of Geology* in the 1830s. In spite of this, palaeontology was established as a science in the early 1800s.

The story of the famous Maastricht mosasaur makes a convenient point at which to start the history of the discovery of the great Mesozoic reptiles. At Maastricht in Holland chalk was excavated from shallow underground mines, and in 1780 the jaws and part of the skull of a gigantic marine animal were found by the quarrymen. They were acquired by a German physician, Dr Hoffman, and identified first as those of a whale and then, correctly, by a Dutch anatomist Adrien Camper, as those of a gigantic lizard; the jaws are 4 feet (1.2 m) long. Their fame soon reached the ears of the Canon of Maastricht, Dr Godin, who owned the land overlying the subterranean quarry, and on these grounds claimed the fossil as his own. Hoffman appealed to the law but lost his case, and the Canon took possession of the skull and put it on exhibition in his private museum.

In 1795 the town was besieged by the Republican French army and Dr Godin noticed that, in spite of its proximity to the fort, no shots were being directed by the French artillery at his museum. He immediately suspected that they had instructions to avoid risk of damage to his famous and precious fossil, having designs on it as a trophy of war, so he hid it elsewhere in the town. His suspicions were well founded. As soon as the town surrendered, and the fossil was found to be missing, a scientifically inclined French officer persuaded the commanding officer to offer a reward of six hundred bottles of wine for it. Such a large and conspicuous specimen could not remain hidden for long under this sort of pressure, and the reward was claimed by a happy party of grenadiers the next day.

The fossil was removed to Paris and an account was published by the French officer whose zeal had secured it for his country, describing it as a crocodile. A few years later it was studied by the Baron Cuvier, who confirmed Camper's opinion that the jaws were those of a lizard. For some reason the Baron did not name it, and it was not until 1828 that an English naturalist, the Rev. W.D.Conybeare, bestowed on it the name *Mosasaurus camperi*, after Adrien Camper and the River Maas (Meuse in French, Mosa in Latin) near which it was found.

Cuvier was a very remarkable man. He was a distinguished administrator, serving under the Emperor Napoleon in the Ministry of the Interior, and he also introduced order into the study of fossil and living animals. He made an exhaustive examination of fossil elephants and demonstrated for the first time that whole faunas of animals had lived in the past and become extinct, and he was the authority to whom any problematical vertebrate fossil was submitted.

Dinosaurs were first discovered in the 1820s by two English naturalists working independently of each other. William Buckland successfully combined two careers. He was an eminent ecclesiastic and became Dean of Westminster, and he also held academic posts in mineralogy and geology at Oxford University. In 1824 he published a description, based on fossils from the Middle Jurassic Stonesfield Slate, of a huge carnivorous reptile, for which he and Conybeare devised the name *Megalosaurus* or 'giant lizard'. One of the fossils consisted of part of a jaw, with large blade-like teeth set in sockets. This was the first dinosaur to receive a name.

Two years earlier, in 1822, Dr Gideon Mantell, a physician in the county of Sussex in southern England, was driving out to visit a patient with his wife, who was there for the ride. While he

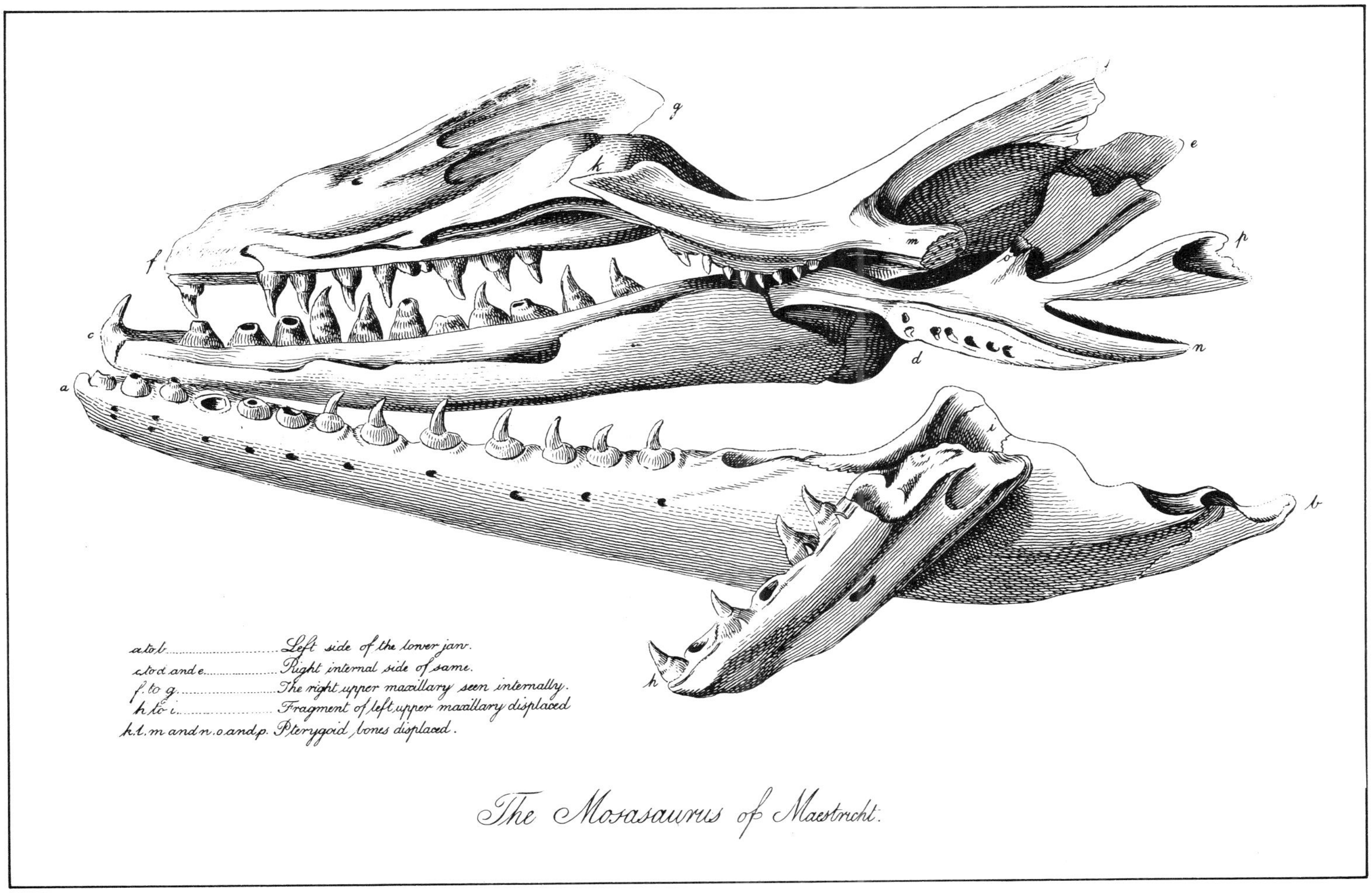

A drawing of the jaws and part of the skull of *Mosasaurus* found in a chalk mine at Maastricht in 1780.

was in his patient's house Mrs Mantell found, in a heap of stones intended for road repairs, a rock with some large and curious teeth embedded in it. Her husband was delighted with them; he was much more interested in fossils than in his medical practice. He located the quarry from which the road metal had come and obtained some more specimens. Although their form and mode of wear indicated a herbivore, Mantell suspected from the first that they were reptilian.

Cuvier saw them and attributed them to a rhinoceros, maintaining that they must have come from a superficial deposit and not, as Mantell maintained, from the Lower Cretaceous Wealden strata. By a lucky chance he met a zoologist who was working on Central American lizards, and a close resemblance was revealed, apart from size, between the fossil teeth and those of an iguana, which is a herbivorous lizard. The obliging Rev. Conybeare suggested the name under which, in 1825, a description of the teeth was published: *Iguanodon* or 'iguana-tooth'. Cuvier very handsomely admitted his error, and in this way a second dinosaur came to be known. Mantell continued to collect and describe fossils of *Iguanodon* and other Cretaceous reptiles. His best specimen came from a quarry near Maidstone in Kent, and a rather whimsical consequence of this is that an iguanodon stands opposite a heraldic lion on the coat of arms of the Borough of Maidstone.

The next personality to feature in the dinosaur story was Richard Owen, who was the originator of the term 'dinosaur'. He was not an explorer after fossils but an interpreter of them, an anatomist who followed in Cuvier's footsteps. He was the first superintendent of the natural history section of the British Museum, and was the moving spirit behind the building of the present Natural History Museum at South Kensington. He turned his attention to the great Mesozoic reptiles in 1841, and proposed the name 'Dinosauria' ('terrible lizards') to include *Megalosaurus*, *Iguanodon*, and another discovery of Mantell's, *Hylaeosaurus* (an ankylosaur). This was, and still is, a collective term; there is no genus 'Dinosaurus'. It was Owen's misfortune that he was too set in his ways and beliefs to accept the theory of evolution, as expounded by Darwin when Owen was already in his fifties. As a consequence he came into conflict with the dynamic advocate of Darwin's views, T.H.Huxley, and suffered humiliating defeats in the academic arena.

OVERLEAF
The small ceratopsian *Protoceratops* lived in the region that is now central Asia. Fossils of eggs, hatchlings and adults have been found together and an adult is seen here watching two young ones emerge from their egg shells.

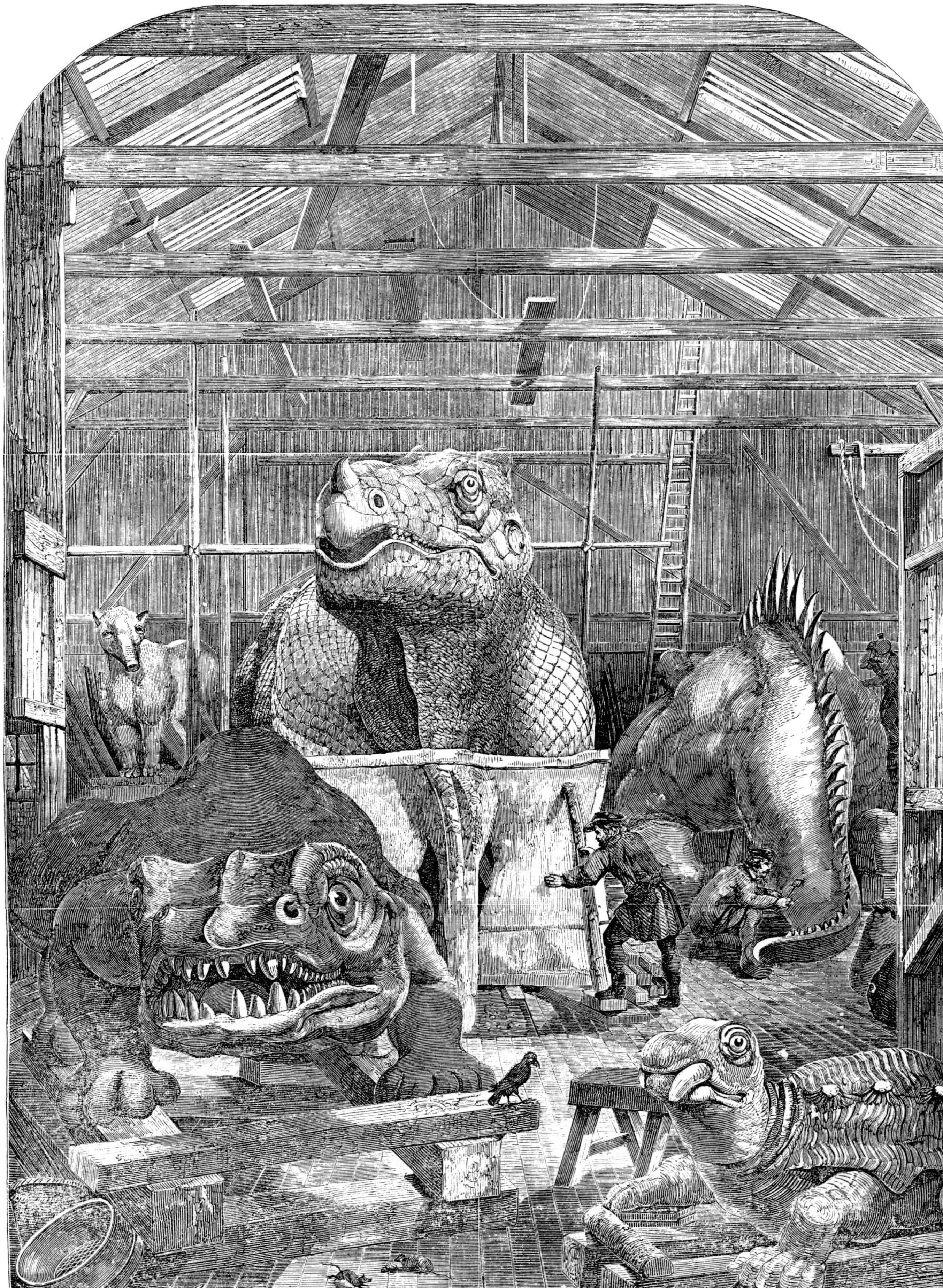

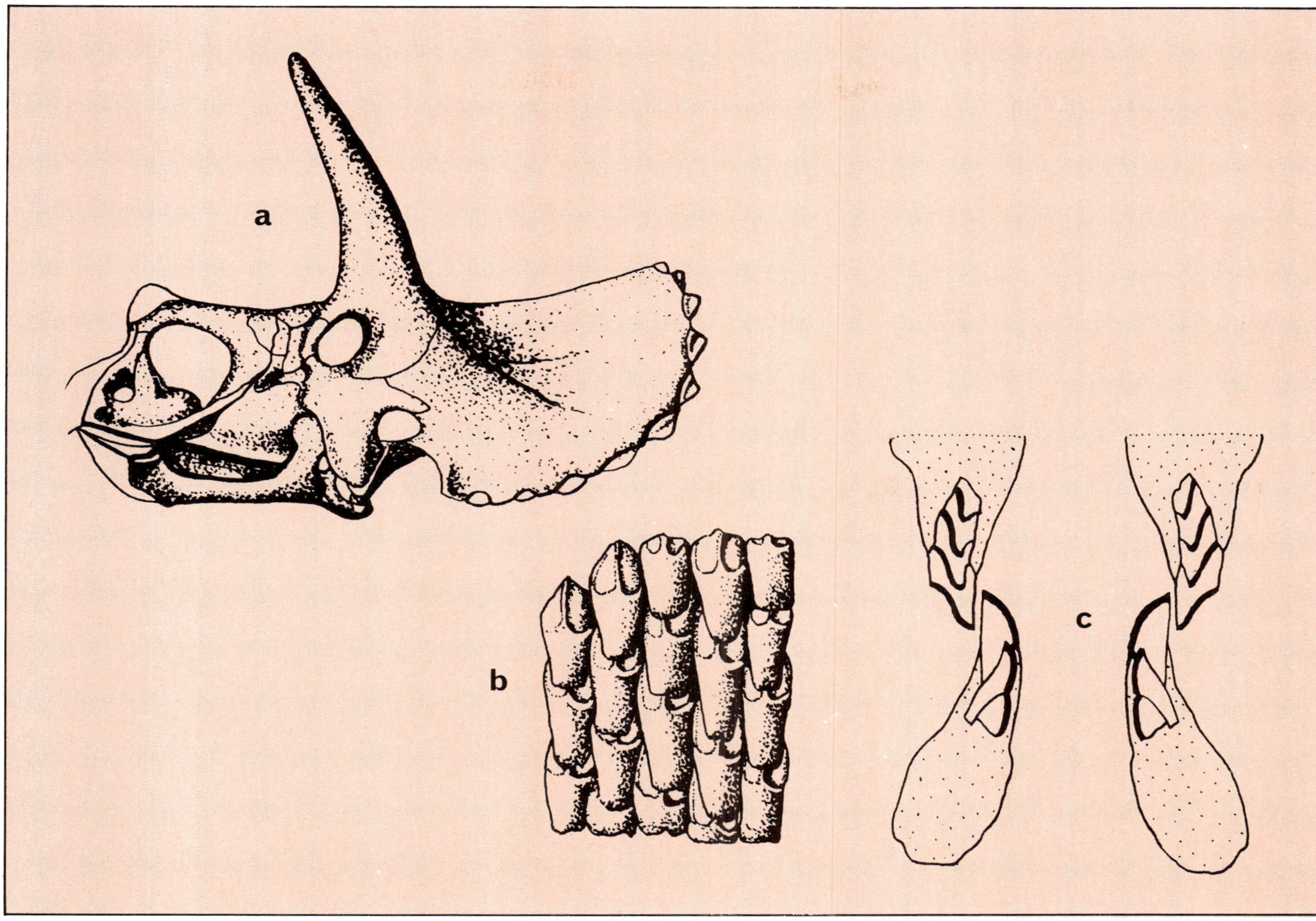

Above Fossil remains of *Triceratops*. (a) The skull. (b) Part of the dental battery showing replacement teeth underlying those that are functioning, which will fall out when they become worn. (c) Diagram to show the shearing action resulting from the teeth of the lower jaw sliding inside those of the upper jaw.

Opposite A contemporary impression of the workshop at Crystal Palace where Waterhouse Hawkins constructed his dinosaur models.

On the fringes of the early dinosaurian research was the slightly fantastic figure of Mr Waterhouse Hawkins. He was no academic but a blend of sculptor, modeller and builder, and he constructed the famous dinosaur models that enlivened the Crystal Palace Exhibition when it was moved out of London in 1854 to Sydenham, where they can still be seen. They were made under Owen's supervision, and it is recorded that the materials used in the iguanodon model comprised: 600 bricks, 650 two-inch half-round drain-tiles, 900 plain tiles, 38 casks of cement, 90 casks of broken stone, 100 feet of iron hooping and 20 feet of cube-inch bar. Shortly before its completion Hawkins invited the leading scientists of the time to a dinner party inside the iguanodon on 31 December 1853. Twenty-one guests are said to have been present and Professor Owen, at the head of the table, presided over what must have been a rather crowded and chilly occasion.

One more Englishman played an important part in the nineteenth century in advancing our understanding of dinosaurs. This was H.G.Seely, who was the first to see that they could be classified in two distinct Orders based on the structure of the pelvis, and in 1887 he proposed the ordinal names 'Ornithischia' and 'Saurischia', with which readers who have got as far as this will be familiar. Seely also published a treatise on the pterosaurs, called *Dragons of the Air*, which is regarded as a classic of palaeontology.

In North America dinosaur footprints were known early in the nineteenth century; Edward Hitchcock studied and recorded them carefully, but attributed them to birds. It was in the 1850s, when the Victorian English were staring wide-eyed at Hawkins's models, that skeletal fossils were first discovered. These came from the Cretaceous of Nebraska Territory (now Montana) and included teeth of a hadrosaur and of a theropod not unlike *Megalosaurus*. They were submitted to Joseph Leidy of the Academy of Natural Sciences of Philadelphia, who immediately recognized their affinity with the finds made in England. Then a skeleton came to light in a marl-pit near Philadelphia and was excavated with admirable care by W. P. Foulke. The pelvis, arm and leg bones and some teeth and vertebrae were recovered, and Leidy, who was an

OVERLEAF
The ceratopsian dinosaurs of the late Cretaceous Period developed horns and neck-frills in a variety of different patterns. *Monoclonius* had a fairly small frill and only one large horn on its nose. *Styracosaurus* also had a single nasal horn, but its neck-frill was broken up into spikes all round the edge.

excellent anatomist, was able to recognize, from the teeth, an affinity with *Iguanodon*, and from the limbs the fact that this dinosaur was a biped. As Hawkins's models and contemporary drawings show, Owen had pictured both *Iguanodon* and *Megalosaurus* as ponderous quadrupeds, and the credit for standing the ornithopods on their hind legs belongs entirely to Joseph Leidy, who named his dinosaur *Hadrosaurus foulkii*. Eight years later E.D. Cope, a pupil of Leidy, found in the same marl-pits a skeleton of a theropod, very like *Megalosaurus*, well enough preserved to show that it too was a biped, with even greater disparity between the fore and hind limbs than that seen in *Hadrosaurus*; so in 1866 the theropods were raised up on their hind legs as well.

The discovery of *Hadrosaurus* coincided with the American Civil War (1861–65). After hostilities were over who should appear on the scene but Waterhouse Hawkins, with a commission to model the new American dinosaurs. He spent some years in the country but encountered difficulties involving local politics. A number of his full-scale models, intended for exhibition in the New York Central Park, were condemned and unceremoniously buried in the park. It is pleasant to think of archaeologists of some remote future time excavating the city of New York and inventing theories to account for Hawkins's dinosaurs.

At this point we must turn to those two remarkable protagonists of American vertebrate palaeontology, Edward Drinker Cope (the pupil of Leidy mentioned above) and Othniel Charles Marsh. The relationship between the two was of a distinctly 'hare and tortoise' type. To begin with Cope had every advantage. His family was wealthy and he was clever and precocious. At the age of eighteen he published a paper on the classification of salamanders, and thirty more papers come from his hand in the next three years. After a tour of museums in Europe he was appointed to a professorship of zoology at the age of twenty-four. Unfortunately he was hasty, jealous and quarrelsome as well as being clever and energetic, and his whole career was coloured by this combination of qualities.

Marsh came of an obscure farming family and was still uneducated and without prospects at the age of twenty. His good fortune was that his mother's brother was none other than the self-made millionaire George Peabody, who had a high regard for education because he had never enjoyed its advantages. When he was twenty-one Marsh used a settlement derived from his uncle to send himself to school, and after four years there asked Uncle George to give him a university education at Yale, and his request was granted. He was slow but hard-working, and he finished his education at the age of thirty-one well-qualified and high in his uncle's esteem. After a tour to Europe he persuaded his patron to make a generous endowment to Yale for a new museum in which he was to be professor of palaeontology, secure for the rest of his life and with no duties other than research, for which the endowment amply provided. He was nine years older than Cope, but their scientific careers started at about the same time and both pursued identical ends: the discovery and description of extinct vertebrate animals, especially of the great Mesozoic reptiles. Both had plenty of money for costly exploration in the field.

It would be difficult to find two more incompatible temperaments, and there is a story of a hasty reconstruction by Cope of a long-necked plesiosaur in which he somehow got the thing back to front, with the head at the wrong end. Marsh, on a visit to Philadelphia in 1870, spotted the rather ridiculous error and drew Leidy's attention to it. Cope was furious and Marsh unconciliatory, and from then on the two feuded bitterly, both on paper and in the field.

Their first battleground was the famous Upper Jurassic Morrison formation of Colorado and Wyoming. Here their rival working parties made the earliest finds of well-preserved sauropods, *Brontosaurus*, *Camarasaurus* and *Diplodocus*, of *Allosaurus* and *Stegosaurus* and many more besides. Always in competition, they also collected in the Upper Cretaceous of Montana and made spectacular finds of duckbills, ceratopsians and ankylosaurs. Marsh was the inventor of the tissue-paper and plaster-bandage method of extracting bones intact from the rock. Characteristically Marsh continued prosperous to the end of his days while Cope speculated in mining shares and lost his private fortune. There has been speculation as to whether the two would have achieved more if they had worked in amicable collaboration, or whether their rivalry spurred them to efforts greater than they would have made in concert. However this may be, both were great men who devoted their lives to research on the richest sources of dinosaur remains in the world, and their memory deserves respect in equal measure.

During the early twentieth century R.S. Lull became interested in the American Triassic dinosaurs, the authors of the footprints attributed to birds by Hitchcock. In 1922 Lull became director of the Peabody Museum, so that the mantle of Marsh fell on his shoulders.

Rather earlier, in the late 1800s, the American Museum of Natural History in New York acquired an energetic president, Henry Fairfield Osborne. Work on the Wyoming Jurassic localities was resumed under his direction. One of Marsh's localities, known as Como Bluff, was revisited and a new one was found at which a shepherd had built a shelter of the dinosaur bones which were lying about on the surface; here the famous Bone Cabin Quarry was opened. Work continued there up to 1905 and a lot of the material now exhibited in the museum's two large dinosaur halls was obtained there.

Another American institution to play a major role in dinosaur-hunting was the Carnegie Museum at Pittsburg. Its best remembered fieldworker was Earl Douglas, who discovered near Vernal in Utah strata of Jurassic age which had been vertically tilted by earth-movements. Development here has produced the famous Dinosaur National Monument at which a large vertical rock face has been prepared showing skeletons *in situ* of sauropods and other dinosaurs.

Another chapter in the history of research in North America opened in 1910 along the cliff-like banks of the Deer River of Alberta in Canada. Two parties prospected and worked the late Cretaceous strata of this area. One was led by Barnum Brown of the American Museum of Natural History and the other by a Canadian contingent consisting of Charles Sternberg and his three sons. Sternberg had previously worked in Kansas and Wyoming, where two of his sons found the two celebrated 'mummies' of *Anatosaurus*. Brown had had previous experience in Montana, where he excavated the most complete skeletons known of *Tyrannosaurus rex*. On the Deer River Brown had a large flat barge constructed on which the party lived and stored their tons of fossil-containing rock, collected as they slowly floated and moored down river. Later the Sternbergs adopted the same method. The two parties worked in complete harmony and there were plenty of fossils to keep both of them busy. They made spectacular finds of ceratopsians, duckbills and of the huge carnivore *Gorgosaurus*.

One more American foray for dinosaurs must be mentioned, this time far afield in Mongolia. In 1922 and 1923 expeditions were led by Roy Chapman Andrews and the palaeontologist Walter Granger, and they found at a place called Ulan Bator the famous nests, eggs and various growth stages of *Protoceratops*.

Although European dinosaurs were the first to be recognized only one spectacular find of

OVERLEAF
The remarkable primitive bird *Archaeopteryx* is not believed to have been capable of true flight. It probably lived among the trees, gliding between them much as flying squirrels do today.

An impression of the dinner party held in Waterhouse Hawkins's Iguanodon model in 1853.

Right An early restoration of *Iguanodon* and *Megalosaurus* made about 1860. Note the incorrectly placed horn on the *Iguanodon's* nose, and the fact that the animals were thought to be quadrupeds.

Below An early restoration of *Ichthyosaurus* and *Plesiosaurus* made about 1860. The two spouts of water from the *Ichthyosaurus's* nostrils are purely fanciful, and they were not known at that time to have a fish-like tail.

OVERLEAF
Above left A fossil skull of *Triceratops*. The jaw muscles ran right back to the neck-frill and were over a metre long.
Below left The Coat of Arms of the Borough of Maidstone, a town in southern England. The figure on the left is an *Iguanodon*, included to commemorate the fact that the earliest important find of this dinosaur was made near the town.

Above right One of the dinosaur models made by Waterhouse Hawkins in the 1850s, as it now appears in the grounds of the Crystal Palace Park in south London, England.
Below right The Dinosaur National Monument has been created in Utah, in the United States, at a locality where Jurassic strata, containing numerous fossil bones, have been tilted into a vertical position by earth movements. The fossils can now be seen exposed on a wall of rock, a spectacle not available anywhere else in the world.

Below An early restoration of a pterosaur of the Jurassic Period, made about 1860. It is of better quality than most of the restorations of that time.

1549
1949
AGRICULTURE AND COMMERCE

HYLAEOSAURUS

them has been made there. This is of course the Bernissart iguanodons, described in Chapter 3 and discovered in 1878, just after Cope and Marsh had started their work on the Jurassic dinosaurs of the western United States. Credit for recovering the iguanodon material goes largely to M. De Pauw and to the mine officials who gave much assistance during the three years of excavation. Subsequent preparation of the skeletons occupied a quarter of a century and was the life-work of Louis Dollo, who became the world's only iguanodontologist. He established all the details of their anatomy, the thumb-spike, the remarkable ornithopod tooth-succession (seen also in the hadrosaurs) and the presence of a larger and a smaller type of iguanodon. Dollo thought these were distinct species, but they are now regarded as simply males and females.

Triassic dinosaurs were found in Europe early in the present century at Trossingen in south Germany. From here Friedrich von Huene described the prosauropod *Plateosaurus*. Later in his career he went to South America and made a study of late Cretaceous dinosaurs in Argentina and of Triassic ones in Brazil. The best source of Triassic dinosaurs is South Africa, where they were found and exploited quite recently by A. W. Crompton of the Peabody Museum and Alan Charig of the Natural History Museum in London. All our knowledge of Triassic ornithischians comes from this locality.

The wonderful dinosaur graveyard at Tendaguru was discovered in 1907, when the territory that is now Tanzania was a German colony. From 1909 to 1912 excavation was conducted there under the leadership of Werner Janensch of the Berlin Museum. The operation was most efficiently conducted on a very large scale, and about two hundred tons of fossil bones, carefully extracted and encased in plaster, were sent back to Germany. The results were spectacular and included the huge skeleton of *Brachiosaurus* now mounted in the Berlin Museum with its skull 40 feet (12.3 m) above the floor.

The dinosaur expeditions and discoveries of the present century are too numerous for description in a single short chapter. In Central Asia the American lead in 1922 has been followed-up with energy and success. Soviet research in Upper Cretaceous strata discovered *Tarbosaurus*, a close relative of *Tyrannosaurus*, and numerous finds of hadrosaurs and ceratopsians have established the remarkable affinity between the Central Asian and North American Upper Cretaceous dinosaurs. The name of Dr I. Efremov is closely linked with the Soviet expeditions of the 1940s and 1950s. The Polish and Mongolian governments have also made combined expeditions to the same region, starting in 1964. The huge fore limbs of *Deiocheirus* form one of their most spectacular finds. In Kazakhstan, Dr A. G. Sharov of the Soviet Academy of Sciences found in Upper Jurassic strata the wonderful furry pterosaur which he named *Sordes pilosus*, and he was also the discoverer of the pterosaur 'ancestor' *Podopteryx*, found in the Triassic of Kirgizstan.

In North America, the real home of dinosaur science, new discoveries are constantly being made, and work goes on quietly on the enormous backlog of material stored in the museums. Lively minds like those of Edwin Colbert, John Ostrom and Robert Bakker keep the subject before our eyes and remind us that, although we already know enough about dinosaurs to make them a rewarding subject for discussion, we shall know more and more in years to come.

Glossary

Age of reptiles The four successive geological periods, Permian, Triassic, Jurassic and Cretaceous (280–65 million years ago) during which reptiles were the dominant animals on land and in the surface waters of the sea.

Ammonites An extinct group of cephalopods with shells in the form of a flat spiral.

Amphibian A vertebrate animal which lives in water as a 'tadpole' when young, breathing like a fish, and develops into an air-breathing adult.

Amphibious Living partly on land and partly in the water.

Ankylosaurs The armoured dinosaurs, one of the four groups comprising the Ornithischia.

Arachnids The group of arthropods comprising the spiders, scorpions and allied animals.

Archosaurs The 'ruling reptiles', comprising the thecodonts, crocodiles, pterosaurs and the two orders of dinosaurs. The archosaurs dominated or 'ruled' the lands of the earth through most of the Mesozoic era. Birds are descended from archosaurs.

Arthropods The great group or phylum of animals including centipedes, spiders, crabs, insects etc. and characterized by having jointed legs and segmented bodies.

Brachiopods Members of a group of animals with shells rather like those of bivalve molluscs, but not allied to the molluscs.

Carnosaurs The large heavily built members of the theropod dinosaurs.

Cephalopods The most highly developed group of molluscs, including the living octopus, cuttlefish and squid and the extinct ammonites and belemnites.

Ceratopsians The horned dinosaurs, such as *Triceratops*; one of the four groups comprising the Ornithischia.

Class The taxonomic category above Order.

Coelurosaurs The small active members of the theropod dinosaurs.

Cretaceous The latest period (136–65 million years ago) of the Mesozoic era, and the last of the four periods comprising the age of reptiles.

Crustaceans The group of arthropods comprising the crabs, lobsters and allied animals.

Dromaeosaurs A late, highly developed group of small theropod dinosaurs.

Ectothermic 'Cold-blooded', having a variable internal temperature, dependent on external conditions.

Endothermic 'Warm-blooded', having a constant internal temperature independent of external conditions.

Fossil Remains of an animal or plant, or evidence of its presence, preserved naturally from past time.

Genus plural *genera* A taxonomic grouping of closely related species. Dinosaurs are generally referred to by the generic name, e.g. *Tyrannosaurus, Iguanodon.*

Globigerina ooze An accumulation on the sea floor of minute shells of a species of Foraminifera, which are primitive (protozoan) animals with shells.

Gondwanaland The great southern land-mass that existed before the start of continental drift.

Hadrosaurs The duck-billed dinosaurs, such as *Anatosaurus*, included in the ornithopod group.

Ichthyosaurs Marine reptiles of the Mesozoic era, shaped very much like fishes.

Jurassic The middle period (193–136 million years ago) of the Mesozoic era, and the third period of the age of reptiles.

Laurasia The great northern land-mass that existed before the start of continental drift.

Mesozoic The geological era (225–65 million years ago) comprising the three periods Triassic, Jurassic and Cretaceous. Dinosaurs existed from the early part of this era up to the end of it.

Metabolic rate The rate at which energy is produced in the body of an animal by oxidation or burning of its food.

Mollusc A soft-bodied animal usually having a shell; the majority are marine, e.g. snails, oysters, scallops and also octopus and cuttlefish.

Mosasaurs Gigantic marine lizards which existed only during the later part of the Cretaceous period and became extinct at the end of it.

Glossary

Order A major category in the classification of animals; Orders are grouped into Classes and subdivided into suborders or families.

Ornithischian Member of the Order of dinosaurs characterised by a bird-like pelvis. All the ornithischians were herbivorous (see ornithopods, stegosaurs, ankylosaurs, ceratopsians).

Ornithopods One of the four groups of dinosaurs comprising the Ornithischia, and including *Iguanodon* and the hadrosaurs.

Palaeontologist A student of fossils and ancient forms of life.

Pelycosaurs The earliest group of reptiles to become dominant animals on land. They gave rise to the therapsids, which were the ancestors of mammals.

Permian The latest period (280–225 million years ago) of the Palaeozoic era, and the first of the four periods comprising the age of reptiles.

Plesiosaurs Marine reptiles of the Mesozoic era. Some were long-necked and lived at the surface, others were whale-shaped and dived in search of food.

Pterosaurs Flying animals, usually classed as reptiles, of the Mesozoic era. Often called 'pterodactyls'.

Predator An animal that obtains its food by hunting and killing.

Saurischian Member of the Order of dinosaurs characterised by a reptile-like pelvis. Both carnivorous and herbivorous forms are included (see theropods, sauropods).

Sauropods The herbivorous members of the saurischian Order of dinosaurs. Characterised by long necks, long tails and gigantic size, e.g. *Brontosaurus, Diplodocus.*

Species (plural the same as the singular) A particular 'kind' of plant or animal, usually referred by its generic and specific name in combination, e.g. *Tyrannosaurus rex* (see genus).

Stegosaurs One of the four groups of dinosaurs comprising the Ornithischia.

Taxonomy The science of classifying and naming animals and plants.

Thecodonts The early and ancestral group of the archosaurs.

Therapsids The mammal-like reptiles, dominant in late Permian and early Triassic times. Evolutionary ancestors of the mammals.

Theropods The carnivorous members of the saurischian dinosaurs, all bipeds and including *Allosaurus, Compsognathus.*

Triassic The first period (225–193 million years ago) of the Mesozoic era, and the second period of the age of reptiles.

Trilobites A group of primitive arthropods that lived in the sea during the Palaeozoic era.

Vertebrate An animal with a backbone.

Further Reading

CHARIG, A.J. and HORSFIELD, B., 1975. *Before the Ark*. BBC Publications, London.

COLBERT, E.H., 1962. *Dinosaurs: Their Discovery and their World*. Hutchinson, London; Dutton, New York.

COLBERT, 1965. *The Age of Reptiles*. Weidenfeld & Nicolson, London.

COLBERT, 1969. *Evolution of the Vertebrates*, 2nd ed. Wiley, New York.

COLBERT, 1971. *Men and Dinosaurs: The Search in Field and Laboratory*. Penguin.

COLBERT, 1974. *Wandering Lands and Animals*. Hutchinson, London.

DESMOND, A.J., 1975. *The Hot-blooded Dinosaurs*. Blond & Briggs, London.

HALSTEAD, L.B., 1969. *The Pattern of Vertebrate Evolution*. Oliver & Boyd, Edinburgh; Freeman, San Francisco.

HALSTEAD, 1975. *The Evolution and Ecology of the Dinosaurs*. Eurobook, London.

KURTEN, B., 1968. *The Age of Dinosaurs*. Weidenfeld & Nicolson, London.

SEELEY, H.G., 1901. *Dragons of the Air*. Methuen, London (Reprinted 1967, Dover paperback).

SWINTON, W.E., 1965. *Fossil Amphibians and Reptiles*, 4th ed. British Museum (Natural History).

SWINTON, 1965. *Fossil Birds*, 2nd ed. British Museum (Natural History).

SWINTON, 1969. *Dinosaurs*, 4th ed. British Museum (Natural History).

SWINTON, 1970. *The Dinosaurs*. Allen & Unwin, London; Wiley, New York.

Acknowledgements

The original colour illustrations, and the artwork on pages 47, 52 and 60, are by Wilcock Riley Graphic Art.

The illustration on page 114 was taken from original drawings by D. W. Ovendon.

The remaining photographs and illustrations are supplied by, or reproduced by kind permission of the following:
Ardea Photographics: *37* (above & below)
British Museum (Natural History): 12, 16, 65 (above), 109
Courtesy of the American Museum of Natural History: 18, 80, 85 (below), 97, 100
Institut für Palaontologie und Museum der Humbolt-Universität: 34 (above)
Mansell Collection: 65 (below), 129
Natural History Photographic Agency (H. R. Allen): *33* (above and below) *36* (top left),
Natural Science Photos: *36* (below), *36–37* (top centre)
Pat Morris Photographics: 53, *134* (above), *135* (above & below)
Radio Times Hulton Picture Library: 93
Ronan Picture Library: 121, 124
Smithsonian Institute, Washington: 57, 61, 72, 84, 88
Studio Briggs: 92
Tasmin Trenamen: 34 (below), 56, 85 (above) 125

Index

Index

Index